Managing Energy Demand

Technology: Level Two

T206 **ENERGY FOR A SUSTAINABLE FUTURE**

Block 3

Managing Energy Demand

Edited by Godfrey Boyle

The Open University

This publication forms part of an Open University course T206. Details of this and other Open University courses can be obtained from the Course Information and Advice Centre, PO Box 724, The Open University, Milton Keynes MK7 6ZS, United Kingdom: tel. +44 (0)1908 653231, e-mail general-enquiries@open.ac.uk

Alternatively, you may visit the Open University website at http://www.open.ac.uk where you can learn more about the wide range of courses and packs offered at all levels by The Open University.

To purchase a selection of Open University course materials visit the webshop at www.ouw.co.uk, or contact Open University Worldwide, Michael Young Building, Walton Hall, Milton Keynes MK7 6AA, United Kingdom for a brochure. tel. +44 (0)1908 858785; fax +44 (0)1908 858787; e-mail ouwenq@open.ac.uk

The Open University
Walton Hall, Milton Keynes
MK7 6AA

First published 2003

Edited, designed and typeset by The Open University.

Printed in the United Kingdom by Bath Press, Glasgow.

ISBN 0 7492 5393 2

1.1

Contents

The T206 Course Team

Academic Staff

Godfrey Boyle, *Course Team Chair*
David Crabbe
Mike Davies
Dave Elliott
Bob Everett
Ed Murphy
Stephen Potter
Janet Ramage
Robin Roy
Derek Taylor
James Warren

External Assessor

Erik Lysen, *University of Utrecht*

Consultants

Tony Day
Marcus Enoch
Martin Fry
John Garnish
Geoff Hammond
Horace Herring
Ben Lane
Stephen Larkin

Production staff

Jill Alger, *Editor*
Margaret Barnes, *Course Secretary*
Sylvan Bentley, *Picture Researcher*
Philippa Broadbent, *Buyer, Materials Procurement*
Hannah Brunt, *Graphic Designer*
Sam Burke, *QA Engineer*
Daphne Cross, *Assistant Buyer, Materials Procurement*
Alan Dolan, *Course Manager*
Claire Emburey, *Course Secretary*
Clive Fetter, *Editor*
Alison George, *Graphic Design Manager*
David Gosnell, *Software Designer*
Rich Hoyle, *Graphic Designer*
Lori Johnston, *Editor*
Jo Lambert, *Learning Projects Manager*
Katie Meade, *Rights Executive*
Lara Mynors, *Project Manager*
Lynda Oddy, *QA Testing Team Manager*
Jon Owen, *Graphic Artist*
Deana Plummer, *Picture Researcher*
Andy Reilly, *Senior Editor*
Jon Rosewell, *Software Manager, CES*
Karen Ross, *Course Manager*
Mark Thomas, *Deputy WebTeam Manager*
Howard Twiner, *Graphic artist*

BBC staff

Anne Marie Gallen, *Series Producer*
Phil Gauron, *Producer*
Marion O'Meara, *Production Assistant*

Chapter 1

Sustainability, Energy Conservation and Personal Transport

by Stephen Potter

Figure 1.1 Traffic congestion on the Paris Peripherique urban motorway. Traffic congestion is now a growing part of everyday life, not just in city centres but almost everywhere in developed economies. Road building has failed to cut congestion, leading to policies to try to manage the amount of traffic

1.1 Technical and consumption factors in transport's environmental impacts

Not surprisingly, when exploring the issue of energy for a sustainable future, transport has already emerged as an important topic. In a number of places in Book 1, *Energy Systems and Sustainability*, transport was mentioned as a major user of energy and its environmental impacts highlighted. An overview of the transport sector featured towards the end of Book 1 Chapter 1 and the growth in energy use in transport was covered in Chapter 3 (particularly Section 3.3). The growth in the use of oil for transport was covered in Chapter 7, Section 7.6, where it was noted that in the year 2000, 62 % of oil consumption in the UK was for transport (whereas in 1967 it was only 25 %). Transport has come to dominate oil use in the UK and in most other developed countries. Motor vehicle engine technology formed the core of the first half of Chapter 8 on Oil and Gas Engines, and electric transport traction also featured in Chapter 9. This chapter and the three that follow it focus upon the energy use and sustainability issues associated with the growth of the transport sector in the UK, in other developed economies and worldwide. This introductory chapter will set up a framework for exploring different strategic options, with Chapter 2 then examining technological responses. Chapters 3 and 4 will then look at the role that institutions, such as employers, can play in building more sustainable transport behaviour into their operations.

Over the last 40 years, transport in the UK and in other developed economies has moved from being a relatively small consumer of energy to become the largest and fastest expanding energy sector. About eighty percent of transport energy in the UK is consumed by motor vehicles, and three-quarters of that is by cars, so *personal* transport is a major part of overall transport energy use and emissions. Our lives have become increasingly transport-dependent with, among other things, road congestion growing to unprecedentedly frustrating levels. The issue of what should be done to address the transport crisis has become a high profile and highly contentious subject. This is not surprising as car use is now accepted as normal and restrictions upon the 'freedom to drive' are much resented. Yet transport produces major local and global pollutant emissions, with a whole host of other transport-related issues such as accidents and increasingly sedentary lifestyles leading to adverse health effects and obesity, together with social exclusion and the much-publicized problems and cost of congestion.

When global environmental concerns about transport first emerged in the 1980s, the initial reaction was to separate out environmental impacts from all the other issues of high car use. At that time, emissions from industry and production were viewed as the dominant concern, and for vehicles there was an initial emphasis on reducing emissions from vehicle production, with the automotive industry adopting the use of water-based paints, eliminating CFCs and adopting other clean production and pollution abatement technologies. However, by the early 1990s, environmental life cycle analysis had established that the fuel consumed in driving vehicles represents some 90 % of total life cycle energy consumption[1] (for example, see Teufel *et al*, 1993, Hughes, 1993 and Mildenberger and Khare, 2000).

[1] Life Cycle Analysis is explained further in Box 2.4 in Chapter 2 of this book.

This eventually led to a shift in focus from production to product design. Initially, local air pollution concerns resulted in the promotion of emission clean-up technologies, such as catalytic converters for car exhausts. In conjunction with this came moves towards cleaner fuels, including unleaded petrol, low sulphur diesel and the use of 'alternative fuels' such as liquefied petroleum gas (LPG), compressed natural gas (CNG) and electricity in battery vehicles.

The agenda has now moved on from these air quality concerns, with a growing acknowledgement that actions are needed to address global environmental impacts, particularly CO_2 emissions from transport (which was discussed in Chapter 1, Section 1.3 of Book 1). The amount of CO_2 generated by transport in the UK has doubled in the last 25 years and transport is the fastest growing source of all emissions.

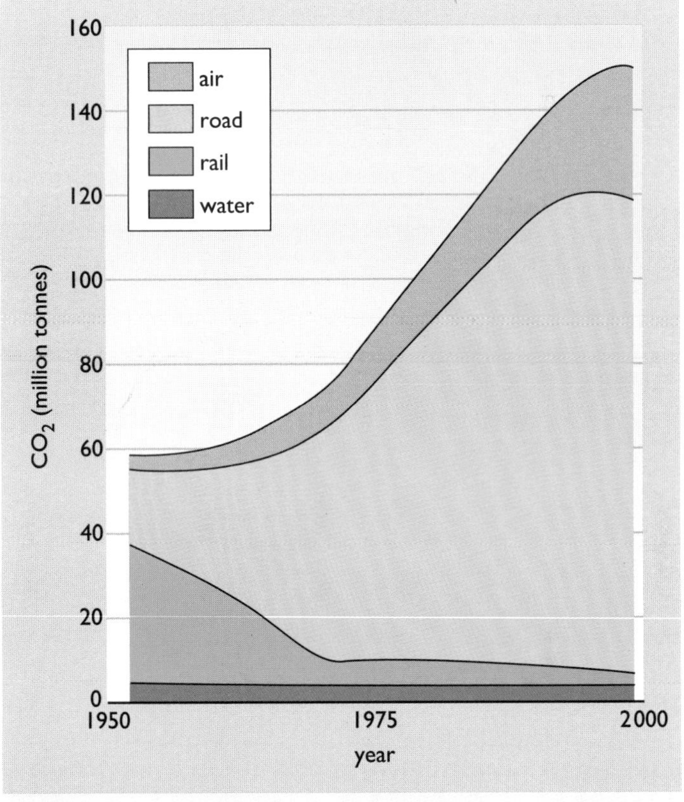

Figure 1.2 UK transport CO_2 emissions 1952–99 (source: Department for transport, UK)

The amount of CO_2 produced when a fuel is burned is basically a function of the mass of fuel consumed and its carbon content. A new emphasis has therefore arisen on fuel type, fuel economy in vehicle designs and the promotion of 'alternative fuels' that have lower carbon content.

In practice vehicle fuel economy improvements may fail to make much of a difference, as increases in the amount of travel, and 'rebound effects', such as changes in drivers' and car buyers' behaviour, compensate for the vehicle improvements. (The rebound effect was discussed at the end of Section 1.5 in Chapter 1). A classic example of such a rebound effect is shown by the 'CAFE' fuel economy regulations in the USA. In the 20 years to the mid 1990s, these *Corporate Average Fuel Economy* regulations improved car fuel economy by over a third, but growing vehicle *use* has more than compensated for this, part of which is due to lower running costs arising from better fuel economy. Overall, although vehicle energy efficiency has improved, the total amount of fuel consumed (and CO_2 emitted) has risen. On their own, product energy efficiency measures do not always save energy, a point which has been made with reference to other energy sectors (see Chapter 5 and Herring, 1999).

The opposite extreme to vehicle efficiency improvements is the consumption-oriented view that behavioural change should be the main policy response to cutting transport's environmental impacts, implying a dramatic reduction in the use of the most energy-intensive transport modes of car and air travel. But if we are to achieve a radical reduction in environmental impacts, many individuals and politicians baulk at the prospect of 'turning the clock back' to a level of mobility considerably less than that we currently enjoy. Car use is now deeply entrenched in our society and economy, however environmentally problematic that may be.

Figure 1.3 The September 2000 Fuel Protests. The political sensitivity of transport was well illustrated by the fuel protests of 2000. When the price of oil led to high fuel prices there were calls for cuts in fuel taxation. Although organized by a very small number of people, the blockade of oil refineries rapidly caused transport chaos and the government quickly caved in, cutting over a billion pounds off fuel and lorry taxes. Environmental and transport policy issues were totally ignored. The cut in tax has subsequently led to a rise in fuel use and CO_2 emissions and the government has amended its CO_2 forecasts accordingly!

1.2 Exploring the issue

Some mix of the two approaches of technical product efficiency improvements and changes in consumption or behaviour would therefore appear appropriate, but what should be the relative contributions of the two? (see Potter, 1998, 2000 and Potter, Enoch and Fergusson, 2001). This chapter therefore seeks to provide an overview of what sort of improvements are needed in product efficiency and changes to consumption patterns to cut CO_2 emissions from personal transport to a sustainable level. Exactly what measures and technologies could be used to achieve these improvements is then considered in detail in Chapters 2, 3 and 4.

A useful way to explore the role of technical and behavioural aspects is to investigate the role of key factors in the generation of transport's environmental impacts. A simple but fruitful approach to such investigations was suggested by Paul and Anne Ehrlich (1990), and developed by Ekins *et al* (1992), in which *Environmental Impact* is expressed mathematically as the product of *Population*, *Consumption* and *Technology*. This formula is:

$$P \times C \times T = E$$

where P is Population, C is the level of Consumption, T is the Technology used and E the resultant total Environmental Impact.

Using this approach, and looking at the world as a whole, an example might be to assume that in the next 50 years or so global population will increase by around 60 % and consumption at least double (see Figure 1.55 in Book 1, Chapter 1). So, if current environmental impacts are expressed as an index of 1.0, then the current or 'baseline' position on the Ehrlich/ Ekins formula is:

$$P \times C \times T = 1$$

If population goes up by 60 %, then its index number would rise to 1.6 and if consumption doubles, its index number becomes 2.0. If the technology does not change (i.e. all energy production and energy use technologies produce the same amount of environmental impacts as today), the formula becomes:

$$1.6 \times 2 \times 1 = 3.2$$

So, if there is no change in the environmental performance of technologies used, overall 'Environmental Impact' (E) increases more than threefold. This is simply a result of more people consuming more goods and resources. To prevent an increase in Environmental Impact, *Technology* (T) has to be reduced to just over 0.3. This might be achieved by, for example, a threefold improvement in energy efficiency or the use of less polluting fuels and technologies – or some combination of the two.

But this is just to stop environmental impacts getting worse! If, for example, a sustainability target suggests we need to halve current environmental impacts then the figure for E has got to be reduced to 0.5. This results in the need for an even bigger improvement in the 'Technology' part of the equation. That would have to go down to 0.16, representing a very large improvement in energy efficiency and/or a major shift to non-fossil fuels.

This is a very simple index model, but it illustrates a crucial point. That is with world population rising and economic growth leading to higher levels of consumption, then very major improvements have to be made in our production and use of energy for there to be any hope of addressing the world's environmental crisis.

What might a transport version of this simple index model look like? Total travel could be broken down into key emission-generating factors which will help explore the role of consumption and vehicle efficiency in cutting environmental impacts from transport. As noted previously, life cycle studies have established that the fuel consumed in driving vehicles represents some 90 % of total life-cycle energy consumption, so this is the issue upon which to concentrate. The Ehrlich-based model can be developed as follows to calculate the environmental impacts from motorized travel. For this, the baseline index of Total Emissions for, say, the year 2003, would be as shown in Table 1.1.

Table 1.1 Baseline Transport Emissions Index

Population	×	Car journeys per person	×	Journey length	×	Vehicle occupancy	×	Emissions per vehicle km	=	Total emissions
1	×	1	×	1	×	1	×	1	=	1

This provides us with a very simple, but nevertheless useful, transport model. It allows us to look at how changes in the values of these five key components will affect the total emissions produced. In this chapter we will use this simple index model to explore possible transport futures and scenarios illustrating different technologies and policy approaches.

1.3 Business as usual

One way to use this index model is to concentrate on the key global issue of CO_2 emissions from personal transport. A reasonable timescale might involve looking ahead 20 years, since much beyond this it is hard to envisage the changes that could occur in transport technologies and policies. So, what level might a transport CO_2 index for the UK reach by the year 2023 if we assume a continuation of current transport trends? With UK population roughly stable, this factor can be left out, but the other key factors are shown in Figure 1.2.

Table 1.2 Key travel trends, based on a continuation of past trends, 2003–23

2003 data and current trends	Index by 2023
Car journeys average about 600 per person per year (currently rising by 14 per year)	1.5
Journey length averages 13.6 km (rising at about 0.14 km a year)	1.2
Car occupancy averages 1.6 (declining by 0.3 % per year)	1.1
Fuel use averages 9.1 litres per 100 km across the UK car fleet. Assumed to improve to 8 litres per 100 km	0.9

Sources: Noble and Potter (1998) and Department for Transport, Transport Statistics Great Britain (editions to 1998)

The rate of fuel economy improvement described in Table 1.2 is faster than that achieved historically (which is only 0.2 % a year). However, the *AutoOil* voluntary agreement between the European Commission and car manufacturers selling in Europe will accelerate this. The AutoOil measure could improve test fuel consumption of the new car fleet by an eventual 25 % in 2012 (see Table 2.4 in Chapter 2 for details). The figure of 8 litres per 100 km is therefore taken as realistic; indeed this is the average level of fuel economy currently achieved in the Netherlands.

For our 20-year timescale, an initial scenario could envisage the continued use of oil for personal transport (the use of alternative fuels will be explored later in another scenario). With the continued use of oil, the carbon content would remain the same, and so emissions would simply be a function of the amount of fuel used. Such a future would result in the formula becoming as shown in Figure 1.3.

Table 1.3 UK Business as Usual (BAU) Transport Emissions Index by 2023

Population	×	Car journeys per person	×	Journey length	×	Vehicle occupancy	×	Emissions per vehicle km	=	Total emissions
1	×	1.5	×	1.2	×	1.1	×	0.9	=	**1.8**

So, if current trends continue, the model indicates that CO_2 emissions will increase to 1.8 times their current level (i.e. a growth of 80 %). This, of course, is only looking at the UK situation. Carbon dioxide emissions and global warming are, however, global issues and an isolationist approach that considers only CO_2 produced in the UK context is inappropriate. Car ownership and traffic levels per capita in the developing world are growing much faster than in the UK. By 2001 there were about 700 million vehicles in the world of which 500 million are cars. These are heavily concentrated in the industrialized nations (15 % of the world's population live in OECD countries, accounting for more than 80 % of car registrations).

Car ownership is forecast to rise sharply in non-OECD countries, particularly in Eastern Europe and Asian economies. If historic rates of growth are maintained, the global car population could exceed one billion in as little as 20 years time with the number of car journeys expected to rise even

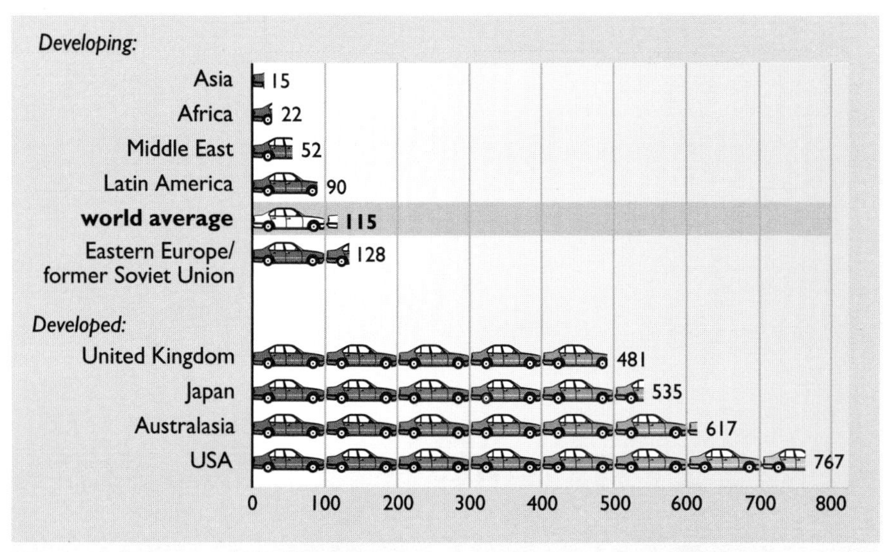

Figure 1.4 Cars per 1000 of population in selected countries and regions, 1995

Table 1.4 Global Business as Usual (BAU) Transport Emissions Index by 2023

Population	×	Car journeys per person	×	Journey length	×	Vehicle occupancy	×	Emissions per vehicle km	=	Total emissions
1.3	×	2.3	×	1.2	×	1.1	×	0.9	=	**3.6**

faster. These growth rates could be incorporated into a global *Business As Usual* (BAU) version of our simple model (Table 1.4).

Population is expected to rise by about 30 % and the growth in car journeys is anticipated to be very large indeed. As noted above, the number of cars in the world is set to double, with the number of journeys expected to rise somewhat faster (hence an index figure of 2.3). Overall, the result of all these trends suggests that CO_2 from personal transport could increase to over three and a half times current levels within 20 years. The indices used are of necessity approximate. Population growth may be less than 30 % in 20 years, but other factors are likely to have been underestimated. The journey length and vehicle occupation figures are as for the UK because global estimates are not available. It is also assumed that fuel economy improves at the UK rate. All of these factors would probably be poorer in developing nations, so if any thing, this underestimates the likely rise in CO_2 emissions, even were population growth to be lower.

The sheer rate of growth in car use in the developing world raises some difficulties totally aside from environmental impacts. As was shown in Book 1 Chapter 7.17 (Figure 7.32), it seems likely that the world production of oil is near its peak. This growth has sustained the massive rise in car, air and freight transport in the developed world. Just as car use is taking off in Eastern Europe and the developing world, oil production is set to peak and start to decline. It is difficult to see how a growth in the numbers of petrol and diesel-engined cars can be maintained for very much longer. Possibly, when oil production fails to meet growing demands, developing countries will be priced out as the developed countries secure their supplies. It is not just environmental impacts and emissions of pollutants that are unsustainable: the long-term availability of oil supplies is also in question. Current growth trends in car use seem to be both economically and environmentally unsustainable, with additional uncertain social, developmental and political implications as well.

This very simple exercise has major implications for any policy to address the CO_2 impacts of personal transport by improvements in vehicle energy efficiency alone. To hold Total CO_2 emissions to their present level, the index figure for CO_2 emissions per vehicle kilometre would have to be drastically cut to compensate for growth in consumption. For the UK, simply to stop Total CO_2 emissions rising would require the emissions index to be cut to 0.56. Expressed in terms of average car fuel economy, this index figure would represent improving average UK car fuel economy from the current figure of 9.1 litres per 100 km to 5.1 litres per 100 km. At the global level, the emissions index would need to be 0.25 to hold transport's CO_2 emissions at current levels, requiring a 4.5 factor improvement in fuel economy so, within 20 years, the world's car fleet would need to average about 2 litres per 100 km.

1.4 Reducing transport's environmental impacts

The thought of achieving a global average car fuel economy of 2 litres per 100 km within 20 years suggests that the sums are starting to look beyond the realms of political (and possibly technical) viability. But this is without even attempting to *reduce* CO_2 emissions from personal transport. Successive reports of the Intergovernmental Panel on Climate Change (for example, Houghton *et al* 1990 and Watson *et al*, 2001) have suggested that a 60 % cut on 1990 levels is needed to mitigate the effects of climate change. Following associated reports by the UK Royal Commission on Environmental Pollution advocating a 60 % cut in UK CO_2 emissions by 2050, with a 40 % cut by 2020, the UK government adopted a target to cut CO_2 emissions by 20 % by 2010. In the 2003 Energy White Paper, the UK government announced a further long-term target to reduce UK CO_2 emissions by 60 % by 2050.

Taking the 2020 target of a 40 % cut, and arbitrarily easing it back to 2023, what sort of efficiency improvements might achieve a 40 % drop in emissions? Returning to the UK, in 2003, transport's CO_2 emissions had already risen by over 15 % since 1990. Therefore, assuming that personal transport needs to make a proportional contribution reducing CO_2 emissions, the index needs to be *not* 0.6 but 0.52. This becomes the target index figure to aim for. Again, if we were to rely on efficiency measures alone, the index for emissions per vehicle kilometre would need to be reduced to around a quarter of current levels (Table 1.5)

Table 1.5 Vehicle Efficiency Improvement required to achieve a 40 % reduction in CO_2 emissions by 2023

Population	×	Car journeys per person	×	Journey length	×	Vehicle occupancy	×	Emissions per vehicle km	=	Total emissions
1	×	1.5	×	1.2	×	1.1	×	0.26	=	**0.52**

So, if fossil fuels were used, and every other trend left unaffected, then fuel economy would need to improve about fourfold, to an average of 2.4 litres per 100 km. Allowing for a proportion of poorer fuel economy vehicles, much of the car fleet would need to achieve under 2 litres per 100 km. Could such an improvement be achieved in 20 years?

The use of smaller-engined, more economical, cars can cut fuel use and CO_2 emissions substantially from present average levels. There are a number of car designs, such as the two-seater *Smart* and some smaller diesel models (see Figure 1.5), that achieve around 3 litres per 100 km. However, their use is not widespread. There is a very big difference between there being some cars that can get close to the target of a fuel economy of less than 2 litres per 100 km and the *whole* car fleet managing that within 20 years. Currently the UK car fleet has an average fuel economy of 9.1 litres per 100 km. Some countries are somewhat better, with Italy, for example, having an average fuel economy of 7.5 litres per 100 km, which is associated with a vehicle tax regime favouring smaller engined cars. The USA averages

Figure 1.5 (a) The two-seater Smart car; (b) there have been a number of other designs like this electric micro car. However the Smart is the only one produced in serious numbers

11.6 litres per 100 km (and there, the increasing use of 4-wheel drive sports utility vehicles is pushing up fuel consumption).

Can car fuel economy be massively improved to achieve a fourfold fuel economy improvement? Highly fuel efficient car designs were mentioned in Book 1, Chapter 1, including the USA 'Hypercar' concept. In the UK, a study by Cousins and Sears (1997) explored what level of fuel economy could be achieved using best practice current technology. This project sought to produce not just a highly fuel-efficient car, but one that would win consumer acceptance. Their study selected a four-seat, five door family hatchback powered by a 600 cc, 23 kW petrol engine, producing a top speed of 152 kph and a performance and price comparable to contemporary small cars (e.g. the Corsa 1.0), but with fuel consumption averaging 2.5 litres per 100 km.

A technology currently being introduced to the market is the hybrid engined-vehicle. These were mentioned in Book 1, Chapter 1 and explained in Section 9.4 of Book 1. By having both electric and internal combustion engines, each type can be utilized at high efficiency. The internal combustion engine is run more constantly, with the electric motor used in slow, stop-start conditions and when strong acceleration is needed. One of the first hybrid cars introduced to the European market, the Toyota Prius, has a test fuel consumption of 4.9 litres per 100 km.

These sort of technologies look like they could deliver an average fuel economy in the 3–5 litres per 100 km range, which is somewhat over halfway to the 20 year target of 2.4 litres per 100 km, if, of course, people were willing to accept such vehicles. Even though the above designs took performance and consumer acceptability into account, people have been very reluctant to buy fuel efficient vehicle. Sales of the Prius and other hybrids are very small, although the Smart has captured a reasonable 'second urban car' niche market. Yet to achieve a 2023 vision of a 3 or 4 litres per 100 km car fleet would require a substantial change in what we think of as a 'car'. It is a future where there would be very few large, heavy or high performing cars. We would probably need to say goodbye to gas-guzzling 4×4 multi-purpose all terrain vehicles, so beloved for use in the urban school run. There would be no room for them in such a high fuel economy future. The vast bulk of the car fleet would have to be modest, low-accelerating vehicles if such good fuel consumption were to be achieved in practice. A key lesson from this simple modelling exercise is that this seemingly 'technical fix' approach would require considerable behavioural change to make it work.

Of course, this ambitious level of fuel economy is referring to developed economies alone. The necessary improvement in fuel economy becomes even greater once a global perspective is taken. This would involve something like the figures shown in Table 1.6.

Table 1.6 Vehicle Efficiency Improvement to achieve Global CO_2 Target

Population	×	Car journeys per person	×	Journey length	×	Vehicle occupancy	×	Emissions per vehicle km	=	Total emissions
1.5	×	2.3	×	1.2	×	1.1	×	0.13	=	**0.52**

Using conventional fossil fuels, in 20 years the global average fuel consumption needs to be just over a tenth of that today – 1.2 litres per 100 km. Taking a longer perspective with further population and car use growth, this would need to improve even more.

1.5 Alternative fuels and renewable energy

For a developed country like the UK, it looks like the widespread use of fuel-efficient vehicle technologies could get us about halfway to a CO_2 emissions reduction target. At a global level, this approach looks far less hopeful as fuel economy improvements, even if achievable, are set to be overwhelmed by a massive increase in consumption. Given such a trend, might the use of less carbon-intensive 'alternative fuels' be the answer? Book 1 and Book 2 looked at ways to introduce low and zero carbon fuels. Carbon intensity (the amount of carbon released in combustion per unit of energy generated) could be added to the index model. This would allow an exploration of the effect of a cut in carbon intensity, for example if vehicle fuel economy remained at the *Business As Usual* rate we started with in Table 1.3. The result of this is shown in Table 1.7, which indicates that a three-fold reduction in carbon intensity would be needed to hit the UK target of a 40 % reduction by 2023.

Table 1.7 Vehicle Efficiency Improvement to achieve Global CO_2 Target

Population	×	Car journeys per person	×	Journey length	×	Vehicle occupancy	×	Fuel per vehicle km	×	Carbon Intensity	=	Total emissions
1	×	1.5	×	1.2	×	1.1	×	0.9	×	0.29	=	**0.52**

Is this possible in 20 years – even if people were willing to accept the type of vehicles concerned? This reduction in carbon intensity is beyond what most alternative fuels can offer. Chapter 2 of this book (Sections 2.6–2.10) will look at alternative vehicle fuels and engines in some detail. Here the emphasis is on the overall environmental performance possible from such fuel changes and to explore the scale of changes needed to achieve such an improvement. A review of studies of the carbon intensity of alternative transport fuels (detailed in Potter, 2000, pp. 65–7) shows that most alternative fuels have been developed to reduce local air pollutants (like the notorious Los Angeles smogs), with little consideration for CO_2 emissions. The first approach is simply to compare petrol and diesel, as diesel is an already widespread alternative to petrol. The CO_2 equivalent lifecycle emissions[2] from diesel are about 25 % lower than from petrol-engined vehicles. Compressed natural gas (CNG) has been in widespread use as a road vehicle fuel in some countries (e.g. Italy and New Zealand) for many years and is emerging as the leading alternative fuel in a number of European states because it enables significant air quality improvements to be achieved. It is also a relatively simple matter to adapt existing engines and designs to use it. However, although CNG produces considerably lower air pollutant levels than petrol or diesel, in terms of climate change gas emissions the reduction is marginal. CO_2 from CNG vehicles is only about 15 % lower than for petrol.

Figure 1.6 A vehicle powered by liquefied petroleum gas (LPG). Like CNG, LPG cuts local air pollution but CO_2 emissions are only marginally less than for conventional road fuels

If electrically-powered vehicles are used, the effect on CO_2 emissions depends on the primary energy source used to generate the electricity and the efficiency of the generation process. This issue was explored in Books 1 and 2. For the average European mix of electricity-generating fuels, an electric car achieves about a 40 % cut in the emission of greenhouse gases compared with a petrol car, and a 22 % cut compared with diesel. However, if coal is the source of primary energy, the CO_2 emissions are slightly worse than those from a petrol-engined car. If gas is used for electricity generation, greenhouse emissions from electric vehicles drop by 50 % compared to petrol, while electricity from nuclear and hydro-electric power stations

[2] CO_2 equivalent emissions was explained in Book 1.

produce the greatest improvement (by 85 %). Although nuclear and most renewable energy sources produce little or no CO_2 during electricity generation, CO_2 is produced in building and maintaining the power stations, which is reflected in these fuel life cycle figures.

The use of Biofuels was emphasized in the 2003 Energy White Paper (DTI, 2003). This anticipated that by 2020 up to 5 % of transport fuels could be biodiesel and bioethanol. However, biofuels are very mixed in their effect on CO_2 emissions. If a car uses an ethanol fuel produced from maize and other crops that are energy-intensive to grow and also energy-intensive to manufacture the fuel itself, there is little or no improvement over fossil fuels (DTI, 2000). However, ethanol from wood comes out well, offering a 66 % cut in CO_2 emissions as the fuel production is less energy intensive. This is also true of biodiesel, produced from rape seed.

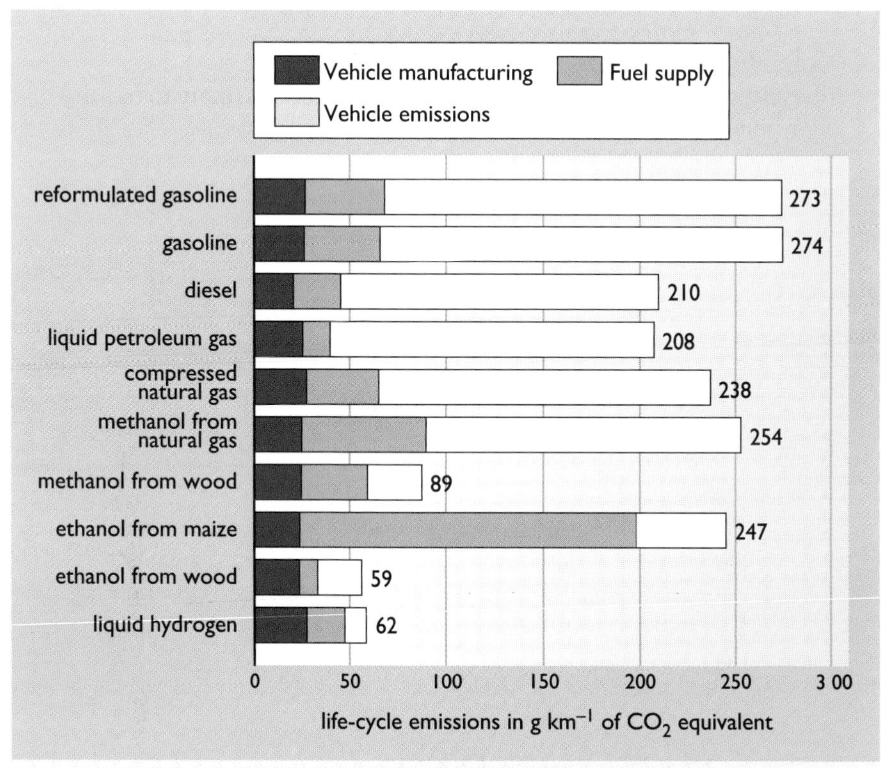

Figure 1.7 Life cycle climate change gas emissions (grams per km of CO_2 equivalent) from alternative fuels

The use of hydrogen fuel cells for automotive applications has attracted a lot of attention, and they look like replacing the battery-electric vehicles as the main challenger to the internal combustion engine (Chapter 2, Section 2.10 cover this issue in some detail). Fuel cells produce electricity using an electrolytic process, converting hydrogen and oxygen to electricity and heat, with water vapour the only emission. But with hydrogen, as with any other manufactured fuel, it is important to take into account the carbon content of any primary fuel used to manufacture it, and other overall lifecycle emissions. Hydrogen produced from renewable energy sources would be a very clean fuel in terms of both local air pollutants and CO_2

emissions. However, at the moment (and for a good while to come) electricity produced from renewable sources seems likely to be used for existing domestic and commercial purposes. It seems unlikely that in a 20 year timeframe there will be enough renewable energy capacity for both existing electricity markets and a new market for hydrogen used in transport.

This issue was explored in a recent report, *Fuelling Road Transport: implications for energy* (Eyre, Fergusson and Mills, 2002). This concluded that until there is a surplus of renewably-generated electricity, it is not beneficial in terms of carbon reduction to use renewable electricity to produce hydrogen for any application, transport or otherwise. The report continues:

> Higher carbon savings will be achieved through displacing electricity from fossil fuel power stations. There would be some carbon savings from hydrogen vehicles using electricity from a power station dependent largely on gas and renewables, if the gas technologies are combined heat and power (CHP). But the supply of hydrogen to mass-market vehicle applications is likely to require more electricity than can be supplied from renewables and CHP alone for at least 30 years.
>
> Eyre, Fergusson and Mills, 2002, p. 4

The situation appears to be that areas other than transport should be prioritized for the use of renewable energy. This report further concludes that the cheapest route to hydrogen would be to produce it from natural gas and that this has 'some potential carbon benefits if used in high efficiency fuel cells vehicles', but that the benefits are relatively small compared to diesel and petrol hybrid vehicles. Overall, the use of alternative fuels for transport presents a very mixed, and extremely uncertain, picture. Over the next 20 years fuel cells seem set to become a mainstream automotive technology, but in terms of CO_2 emissions there may be anything from little effect to a 60 % improvement under very optimistic assumptions. Even the latter does not quite make the 70 % improvement suggested above as necessary were other consumption factors to remain unchanged. In the longer term (30 years or more), hydrogen generated from renewable energy sources would provide an ultimate answer, but we are a long way from having sufficient supplies of renewably generated electricity to achieve this.

Fuel efficiency and alternative fuels - conclusions

This analysis of the role of fuel efficiency and switching to 'alternative fuels' leads to an important conclusion. Even if a purely technical fix, product-level approach were taken, only *a combined strategy of both* improving fuel economy *and* developing alternative fuels stands any hope of getting CO_2 emissions down to a sustainable level. For example, a doubling of fuel economy plus a halving of the carbon content of fuel would hit the UK target. This is potentially achievable, but would require the use of only the very best alternative fuels. At the global level even this approach looks hazardous. Because of the rise in consumption, even if fuel economy doubled, carbon intensity would have to be cut by nearly 80 %, which implies that virtually all primary fuel would have to be nuclear or renewable. Even if fuel economy improved by a factor of four, carbon intensity would still need to be halved (Table 1.8).

Table 1.8 Efficiency and fuel improvement to achieve global CO_2 target

Population	×	Car journeys per person	×	Journey length	×	Vehicle occupancy	×	Fuel per vehicle km	×	Carbon Intensity	=	Total emissions
1.3	×	2.3	×	1.2	×	1.1	×	0.25	×	0.55	=	**0.52**

This scenario is one of highly efficiently produced hydrogen (or other cleaner fuels) powering rather small vehicles. These vehicles would need to have an extremely good fuel economy. The index figure of 0.25 for 'fuel per vehicle' represents the petrol equivalent of 2.3 litres per 100 km compared to 9.1 today. This seems pretty unlikely to be achieved in 20 years. It is hard to envisage that this could be achieved politically, even though it may be just about technically possible. Added to this, beyond 2023, further cuts in CO_2 are required. The IPCC target is for an eventual cut of at least a 60 % in global emissions compared to 1990, which the UK government has now set as a long term domestic target in the 2003 Energy White Paper. Added to this the growth in car use and population will continue, so overall consumption levels will rise, counterbalancing any individual vehicle improvements. As time passes the goalposts move and the whole situation becomes even more challenging.

A 60 % cut in CO_2 compared to 1990 would require the end 'Global Emissions' index figure to be cut to an index value of 0.34. It is hard to speculate with any accuracy how the key factors in the index model will have changed by that date, as it is simply so far ahead. You can put your estimates into the index model and work out what the carbon intensity figure needs to be to hit the 0.34 target for total emissions. My own workings suggest carbon intensity would need to drop to 0.15 or less. What does such a figure mean? One way this index figure could be achieved is if each car uses a quarter of the energy of those around today and gets only an average of 15 % of its energy from fossil fuels. The other 85 % would have to come from renewable sources. The overall result (allowing for the higher energy efficiency as well) is that the average car in 2050 will need to run on under 4 % of the fossil fuels used by the average car today.

Looking this far ahead it could be argued that by then a global, totally carbon-free energy supply system will have evolved. However this exercise does show the magnitude of the challenge ahead. Returning to our more comprehensible 20-year timescale, a very strong technical fix approach might just about achieve the 20-year target at the level of a developed economy like the UK. At the global level, the necessary improvements in fuel economy and type of fuel, even for a 20-year timescale, look unrealistic. Once a longer timescale is envisaged, the whole situation is far more uncertain.

It is also important to add to this the point that our analysis has only been looking at emissions and energy use arising from transport activities.

As was briefly mentioned at the beginning of this chapter, there are a number of important transport policy issues that would be unaffected by using technical measures to reduce emissions. These include traffic accidents, traffic congestion and the host of health related issues linked to sedentary car-oriented lifestyles. All these issues are about the amount of motor traffic rather than how it is powered.

1.6 Travel mode and volume of travel

The magnitude of the changes required in only 20 years looks daunting when a purely 'technical fix' approach to cutting emissions from transport is taken. So, can changes to the 'consumption' elements in our simple index model suggest a more viable path? This would also mean that other issues relating to the volume of traffic would be addressed, such as congestion, accidents and adverse health effects.

In the index model, consumption aspects are expressed in terms of the number and length of journeys and also the index figure for car occupancy. A much-advocated approach is to cut transport's environmental impacts by somehow shifting trips ('modal shift') from the car to less energy-intensive forms of transport. To evaluate this option requires a return to a UK focus, as it is difficult to obtain and use global figures for key factors such as the share of travel by each mode of transport (modal share) and journey length.

A number of studies (detailed in Potter, 2003) have compiled empirical information on the quantities of energy and CO_2 emissions arising from the operations of various transport modes. Table 1.9 is a compilation from these sources for a range of urban public transport vehicles. These figures cover the entire fuel life cycle, allowing for the different engine efficiencies and fuel production systems and the differing carbon contents of the fuels concerned. Clearly, the energy use figures quoted depend very much upon the individual design of vehicles and where and how they operate. The above studies (Potter, 2003) do note variations.

The information for buses was provided by a number of UK urban bus companies and that for railways by London suburban rail operators. The light rail figures were provided for the modern tram operations in Manchester and the metro/underground figure is for the London Underground. The data have been compared with other UK and European studies of energy and CO_2 emissions (CEC, 1992, Climate Care, 2000 and Bestfootfoward, 2000). This comparison suggests that the energy use and CO_2 emission figures for buses and diesel trains are broadly similar to those found in other developed countries. For electric trains, light rail and metros, the energy-use figures are also broadly similar to the UK figures, but CO_2 emissions will vary according to the primary fuel mix of the power stations. The 2000 UK mix of gas, coal and nuclear generation was estimated to produce 480 grams of CO_2 /kWh.[3]

[3] Modern coal power stations produce about 950 grams of CO_2 per kWh of electricity generated and gas combined cycle stations about 450 grams of CO_2 per kWh of electricity generated (Everett and Alexander, 2000). The UK average also includes oil, hydro and nuclear generation.

Table 1.9 Fuel life cycle energy consumption and CO_2 emissions for major transport modes

Mode	Seats	MJ per vehicle kilometre	Kilograms CO_2 per vehicle kilometre	MJ per seat kilometre	Grams CO_2 per seat kilometre
Electric Train	300	117	11.7	0.39	3
Diesel Train	146	74	8.8	0.50	0
Light Rail	265	47	10.1	0.18	38
Metro/ Underground	555	122	26.0	0.22	46
Single Deck Bus	49	14.2	1.6	0.29	33
Double Deck Bus	74	16.2	1.9	0.22	26
Minibus	20	7.1	0.8	0.36	40
Medium-sized Car	5	3.5	0.39	0.70	78

Based upon Carpenter (1994), Potter (2000) and Roy, Potter and Yarrow (2002). See also Fig. 1.49 of Book 1 Chapter 1

In general, the slower forms of public transport using lighter vehicles consume less energy and produce the least CO_2 emissions; however, the Metro/Underground figure seems surprisingly high. There are two reasons for this. One is that it involves an older system (London) and older vehicles than the others for which data were gathered. There is also a problem with urban transport vehicles designed to accommodate standing as well as seated passengers, which particularly affects metro services like the London Underground. Dividing energy and CO_2 emissions by seats thus gives the impression that such vehicles are less efficient than they actually are. For urban and transport policy comparisons, figures for a medium-sized car are also included. This shows that, per seat kilometre, public transport uses a half or less of the energy and generates a half or less of the CO_2 emissions of cars.

To explore the effect of modal shift requires the formula model to be split into the three main components of motorized travel: car, bus and train. This is not to say that non-motorized travel (walk and cycle) is unimportant, but it does not generate CO_2 to any significant extent. Trip shifting to walk and cycle can be accommodated in the model by cutting the 'journeys per person' figure for the motorized modes.

Table 1.10 Baseline CO_2 emissions index for car, bus and rail (2003)

	Journeys per person	×	Journey length	×	Vehicle occupancy	×	Energy use per passenger km	×	Carbon Intensity	×	Modal share
Car	1.0	×	1.0	×	1.0	×	1.1	×	1.0	×	0.88
Bus	1.0	×	1.0	×	1.0	×	0.5	×	1.0	×	0.10
Rail	1.0	×	1.0	×	1.0	×	0.6	×	1.0	×	0.02
Total = 1.02*											

* Not exactly 1.0, but as close as calculations using one decimal point can manage.

The Baseline Index is now as shown in Table 1.10. In this it is taken that the energy use per passenger kilometre by train and bus is, on average, about half that of cars. This ratio is based upon the information discussed above. It could be argued that the energy efficiency of public transport may be somewhat better, but a halving is viewed as a safe estimate. The carbon intensity is similar for all three as oil is the main fuel used for all transport.

A Modal Shift Scenario could be based around the targets suggested by the UK's Royal Commission on Environmental Pollution (1994), which have been used widely in transport policy development. How might these targets be achieved? Some practical examples of measures will be considered in Chapters 3 and 4. Pricing measures, although far from popular, are effective. The introduction in February 2003 of a £5 charge to travel within Central London cut traffic levels by over 20 %. London is the latest of a number of cities to introduce such a scheme. The first was Singapore in 1975, where traffic levels were reduced in a similar way to London and, with regular adjustments to the charging system, road traffic has been held at that lower level ever since. Such pricing schemes often require technological innovation as well. The London congestion charging scheme, for example, is operated through a network of number plate recognition cameras that can distinguish (by links to computer databases) between motorists who have, and have not paid the charge and also identifies those who are exempt. The system can also allow for temporary exemptions, such as an accident leading to traffic being diverted through the congestion charging zone, and ensuring those vehicles are not fined for non-payment. Exemptions to London's congestion charge, incidentally, include alternative fuel cars, so this behavioural change mechanism also stimulates a technological response as well (London sales of hybrid cars are relatively high).

Figure 1.8 The London Congestion Charging Zone

In the following version of the index model it is assumed that, over 20 years, pricing and a whole variety of other modal shift measures will have cut the car's share from 88 % of motorized trips to 65 %, with the bus share increasing to 25 % and rail's to 10 %. In order to show what this can do alone, no technical improvement measures are included. Thus changes to fuel economy are at the 'Business as Usual' historic rate, with the continuing use of oil-based fuels resulting in no change to carbon intensity. The changes in vehicle occupancy would also be at the historic rate, although the shift to public transport is likely to improve average occupancy levels of buses and trains.

Table 1.11 Modal shift and CO_2 emissions scenario

	Journeys per person	Journey length	Vehicle occupancy	Energy use per passenger km	Carbon Intensity	Modal share	Total emissions
Car	1.5 ×	1.2 ×	1.1×	1.1 × 0.88	1.0×	0.65	1.25
Bus	1.5 ×	1.2×	0.8×	0.5 × 0.88	1.0×	0.25	0.16
Rail	1.5 ×	1.2×	0.8×	0.6 × 0.88	1.0×	0.10	0.08
						Overall total = 1.49	

The net result, surprisingly, is a near 50 % increase in CO_2 emissions. This may be better than the baseline 70 % rise in CO_2 without modal shift, as considered earlier, but the cut in CO_2 arising from modal shifts is insufficient to counterbalance the rise in other behavioural factors in the model. An important component of this is trip lengthening, which involves not only motorized trips becoming longer, but also the substitution of short trips on foot by longer trips by car, which is reflected in the rise in the number of journeys per person. Simply to get the total in the index model to equal 1.0 would require the very unlikely combination of the car modal share being cut to 30 %, with the bus share rising to 40 % and the train share to 30 %. Even this would only hold CO_2 emissions at their current unsustainable level. This simple exercise leads to an important conclusion. Not only will technical fix not work on its own, but modal shift, as an isolated policy, is also doomed to failure as a CO_2 reduction measure.

1.7 A multiple approach

The simple index model demonstrates clearly that the only technically (and certainly politically) practical way in which transport's CO_2 emissions can be cut to sustainable levels is to combine changes in *both* the vehicle technology (fuel efficiency and fuel type) and all behavioural factors. Importantly, behavioural change cannot just involve modal shift between different forms of motorized transport. Behavioural change needs to involve a reduction in the trend of trip lengthening and the effect this has on non-motorized travel. A major factor in the increase in road traffic in recent years has been because we make longer trips.

One variation of the index formula that would hit the 40 % reduction target is shown in Table 1.12.

Table 1.12 A multiple approach to achieve UK CO_2 target

	Journeys per person	Journey length	Vehicle occupancy	Energy use per passenger km	Carbon Intensity	Modal share	Total emissions
Car	1.3 ×	1.1 ×	1.0×	1.1 × 0.5	0.8×	0.65	0.41
Bus	1.3 ×	1.1×	0.8×	0.5 × 0.6	0.8×	0.25	0.07
Rail	1.5 ×	1.2×	0.8×	0.6 × 0.6	0.7×	0.10	0.04
						Overall total = 0.52	

This particular combination involves:

- a 30 % increase in car and bus journeys (rather than 50 % in the BAU scenario);
- halving trip lengthening except for rail (assuming this to pick up some long car trips);
- a 50 % improvement in car fuel economy and 40 % for bus and rail;
- a 20 % cut in the carbon intensity of the fuel used for road vehicles and a 30 % cut for rail (the latter probably through electrification).
- Modal shift as in the RCEP report, cutting car from 88 % of motorized trips to 65 %, with bus rising to 25 % and train to 10 %.

The first two factors would involve the proportion of walking and cycling trips being retained or increased, through the use of land use planning policies that reduce the need for motorized travel (through higher densities and less car-based out of town developments).

Overall, for the UK, this means an improvement from our current average fuel economy of 9.1 litres per 100 km to the equivalent of 4.5 litres per 100 km, which is a tough 20 year target, but is probably both technically and politically possible. This improvement in fuel efficiency needs to be combined with the development of alternative fuels to occupy about a third of the market. This also appears to be a tough, but reasonable 20 year aspiration. There would also have to be significant modal shift and a reduction in the rate of trip lengthening to hit the CO_2 reduction target recommended by the scientific community. The number and length of journeys are crucial factors, and yet these are rarely considered in the transport/environment debate.

If all travel generation factors are not addressed, an unrealistic improvement in individual factors is required, as we have explored when looking at technical fix and modal shift options in isolation.

1.8 Reducing transport dependency

The need to reduce the number and length of trips, plus the need to reduce motorized travel, brings in the crucial issue of 'intelligent consumption' with transport systems that deliver the functions of mobility at a lower energy and resource cost. If access to people, facilities and goods can be achieved with less mobility, then this could make an important contribution to cutting transport's environmental impacts. Reducing the need to travel frequently leads into a discussion of land use planning policies and the need to increase urban densities to cut the need to travel. However, this is but one part of reducing transport dependence. Indeed any policy that simply relies on packing people so close together that traffic congestion eventually cuts car use, is probably as doomed to failure as any other single policy measure. There is limited experience of how to travel differently and to enhance accessibility. Some technologies and alternative systems have a potential to reinvent accessibility and mobility in ways that can cut environmental impacts. Again a multiple approach of complementary measures seems appropriate, but this is very much an area of uncertainty, where further understanding is desperately needed.

IT and travel substitution/generation

The travel-substituting potential of the Internet revolution appears, on the face of it, to be strong. I am, at the moment, writing this chapter from home, where I can email colleagues, send and receive documents and have access to a full library and the complete (and overwhelmingly distracting) information resources of the World Wide Web. All this comes to me in my study at the back of my house. In consequence I only go into the Open University's campus two or three times a week. The travel reduction potential of such 'telecommuting' seems obvious; or is it?

A survey of Californian telecommuters by Pendyala *et al.*(1991) provides strong evidence of the positive transport effects of telecommuting. Not only was car use for commuting purposes cut, but non-work trips were also affected. It appears that once telecommuters no longer have a long drive to work, driving long distances for other purposes becomes less acceptable and they tend to undertake shopping and leisure trips more locally.

But there are negative as well as positive 'rebound' effects with this seemingly beneficial technology. Firstly, if telecommuting results in increased energy use in the home, particularly for heating and (in the USA context) air-conditioning, then the overall energy and environmental improvements will be less than envisaged. This again reinforces the need for a life cycle and systems analysis. Of possibly more significance are longer-term lifestyle adjustments to a communications-intensive society. Historically, improvements in the availability and speed of travel have not led people to travel less than they did before, but have always led to lifestyle changes that have resulted in more motorized travel being generated. So, for example, the arrival of buses and trams did not result in people getting to work faster: they moved further away from work and created suburbs. If people need only to travel to a place of work on two or three days a week (or less), this is likely to lead them to live further from work, possibly in more remote locations that are very car dependant for all travel needs (Potter, 1997). In an in-depth analysis of the implications of the 'information society' for spatial planning, Stephen Graham noted that:

> Rather than simply being replaced, transport demands at all scales are rising in parallel with exploding use of telecommunications. Both feed off each other in complex ways, and the shift is towards a highly mobile and communications-intensive society.
>
> Graham and Marvin, 1996

There is a real danger that IT and telecommuting could well result in the generation of more travel than it eliminates.

Reinventing car 'ownership'

An alternative way to manage the use of the car involves not physical or electronic controls, but reinventing the way we obtain and pay for car use. This brings us back to the issue of pricing and economics, which cannot be avoided in transport policy studies. To buy a car involves high fixed costs and once this is made the only relevant costs are those for running it. The most common perception is that fuel is the only cost of a car journey. For public transport, the cost structure is different. There are no separate 'capital' or 'running' costs; all the costs are combined into the price of a ticket. This different way of paying for travel stacks the odds against the bus and train, and, with most of car costs being fixed, there is also little disincentive when trip lengthening occurs. When making an individual journey, it is usual for a car user to compare the fuel costs of travelling by car with the price of a bus or rail ticket. If paying for car use were different, and (as for public transport) the capital and other fixed costs were included in a 'pay by the kilometre' charge, then it is likely that perceptions of the relative cost of car and public transport, and also of short as opposed to longer car journeys, would be different.

One example of this is the *Car Club* concept, in which a fleet of cars is available to individuals who pay for all costs by the kilometre. Car sharing clubs are most widespread in Switzerland, the Netherlands and in parts of Germany, where studies have shown how they affect people's travel patterns. In Switzerland, Harms and Truffer (1999) concluded that car sharing reduced the distances travelled by car. Their research looked at both former car owners who joined the 'Mobility' car club and former non-car owners. Before they joined the car club, former car owners drove less

Figure 1.9 A car and user of the Swiss Mobility Car Club. Half the swiss population lives within a 10 minute walk of a Mobility car park. The inset shows the screen-mounted smart card reader that gives the user access to the car they have booked.

than average. They covered about 9300 kilometres per year by car compared with a Swiss average of 13 000 km. This is to be expected, as people who did not drive much might feel that the fixed cost of car ownership was a lot compared with their limited use of a car. These people would be most attracted by a car club scheme. Even though these people already drove relatively little by Swiss standards, after they became a car-sharing member, this was reduced to 2600 kilometres a year, which is less than 28 % of the distance they previously drove.

Some of the reduction in car travel for Mobility members involved a shift to public transport, bike or motorbike. This accounted for 4000 km, which is about 60 % of the reduction in car use. Significantly, the other 40 % was produced by people cutting trip lengths or finding another way to do things that did not involve travelling at all. Surprisingly, former households without cars did not drive more after they joined the car club. It appears that most of them already had some access to borrowing or hiring cars and the car club was simply a better or cheaper way to carry on doing this.

Harms and Truffer emphasize that the changes in mobility patterns should not be totally attributed to the car-sharing system. Joining a car club was sometimes associated with other changes in peoples' lives, such as moving to another town with different conditions for private and public transport, getting a new job in a different place or with different working conditions, or changes in income. Transport factors did play a certain role. In some cases joining the car club was triggered by a terminal breakdown of their own car or by increasing difficulties with parking, congestion or repair costs of their car.

Overall, maturing car club schemes do suggest that changing the way cars are paid for can have a significant effect upon the mode of travel used, the distance people travel and whether travelling is seen as necessary at all. Forms of obtaining access to cars like car clubs could be developed, particularly if the taxation system were to favour them. However the tax system could also produce a 'car club' effect even for continued private car ownership. Ubbels, Rietveld and Peeters (2002) explored the impacts on car use and the environment of replacing existing taxation on cars and fuel with a kilometre charge for using roads. The redistribution of fixed taxes to a kilometre charge resulted in a modelled reduction in car kilometres travelled of between 18 % and 35 % compared to the base case. CO_2 emissions from cars were cut by 22–40 % and NO_x by 40–50 %. Total travel declined by only 5–10 %, but interestingly public transport travel increased by only a maximum of 5 %. The main impact of the kilometre charge was to increase walking and cycle use by 5–10 % and to increase car occupancy.

The effects of taxation and institutional changes such as these on the ways in which cars use is obtained and paid for would be reflected in the part of our index model relating to the number of trips and their length. It would also result in some modal shift effects as well.

1.9 Conclusions: Travelling lightly

This chapter has explored a framework for thinking through how the personal transport sector could achieve a sustainable level of CO_2 emissions that would meet climate change targets in the medium to long term. It has shown clearly that it is necessary to address all factors generating the overall volume and emissions from the personal transport sector. These include:

- fuel efficiency of vehicle involved
- carbon content of fuels used
- number of journeys made
- journey length
- vehicle occupation
- mode of transport.

Such a multiple approach requires a good understanding of how these factors interact as a system.

In general, technical measures are more understood than behavioural change and trip reduction concepts, and tend to be favoured politically. However, the limitations of the technical fix approach tend not to be appreciated. Behavioural consumption policies are much talked of, but are rarely applied to an effective extent and generally fail to address the full range of consumption factors involved. In particular, approaches to reduce transport dependence appear to offer much potential, but are rarely considered.

Added to all this, there is a serious issue of the differences in timing between the technical fix approach and the 'intelligent consumption', behavioural change approach. Some technical fix measures yield results more quickly than policies to effect change in travel behaviour. Thus a sensible approach would be to use the time that technical product and fuel change improvements can buy to put in place the longer term 'intelligent consumption' behavioural change policies that will 'kick in' as the vehicle and fuel improvement start to run out of steam. The political danger is that technical fixes, being seen as less politically sensitive, will be used to continually put off taking serious behavioural change actions until it is too late.

References

Best Foot Forward (2000) http://www.bestfootforward.com [accessed 6 April 2003].

Carpenter, T. G. (1994) *The Environmental Impact of Railways,* John Wiley.

Climate Care (2000) http://www.co2.org. [accessed 29 May 2003]

Commission for the European Communities (1992) *Green Paper: The Impact of transport on the environment: a community strategy for 'sustainable mobility'*, Luxembourg, CEC Office for Official Publications.

Cousins, S. and Sears, K. (1997) E-Auto: the design of a 2.5l/100 km (113 mpg) environmental car using known technology. *Automobile Environmental Impact and Safety*, ImechE, pp. 369–79.

Department of Trade and Industry (DTI) (2000) *The Report of the Alternative Fuels Group of the Cleaner Vehicle Task Force Report,* DTI, Automotive Directorate, London, The Stationery Office.

Department of Trade and Industry (2003) *Energy White Paper*, London, The Stationery Office (also at www.dti.gov.uk/energy/whitepaper/index.shtml#wp) [accessed 6 April 2003].

Department for Transport. Transport Statistics Great Britain, London, The Stationery Office (annual).

Ehrlich, P. and Ehrlich, A. (1990) *The Population Explosion*, New York, Simon and Schuster,.

Ekins, P. *et al* (1992) *Wealth Beyond all Measure: an atlas of new economics,* London, Gaia,.

Eyre, N., Fergusson, M. and Mills, R. (2002) *Fuelling Road Transport: implications for energy policy*, London, Energy Savings Trust.

Everett, R. and Alexander G. (2000) 'Energy File Part 1: Energy and its use', *T172 Working with our environment: technology for a sustainable future*, Milton Keynes, The Open University.

Graham, S., and Marvin S. (1996) *Telecommunications and the City: Electronic Spaces, Urban Places*. London, Routledge, p. 296

Harms, S. and Truffer, B. (1999) 'Car sharing as a socio-technical learning system: emergence and development of the Swiss car-sharing organisation', Mobility, www.ecoplan.org [accessed 6 April 2003].

Herring, H. (1999) 'Does energy efficiency save energy? The debate and its consequences', *Applied Energy*; 63, pp. 209–26.

Houghton, J. Y. *et al.* (eds) (1990) *Climate Change*, Cambridge, Cambridge University Press.

Hughes, P. (1993) *Personal Transport and the Greenhouse Effect*, London, Earthscan.

Mildenberger, U. and Khare, A. (2000): 'Planning for an environment-friendly car', *Technovation,* 20, pp. 205–214.

Noble, B. and Potter, S. (1998) 'Travel patterns and journey purpose', *Transport Trends*, London, Department of Transport, Environment and the Regions; 1, pp. 3–14.

Pendyala, R., Goulias, K., and Kitamura, R. (1991) *Impact of telecommuting on Spatial and Temporal Patterns of Household Travel: an assessment for the state of California*, Davis, Institute of Transportation Studies, University of California.

Potter, S. (1997) 'Telematics and transport policy: making the connection', in Droege, P. (ed.) *Intelligent Cities*, Amsterdam, Elsevier.

Potter, S. (1998) 'Achieving a factor 10 improvement', in Daleus, L. and Schwartz, B., *F_retag I Kretslopp*, Stockholm, Swedish Energy Agency, pp. 219–26.

Potter, S. (2000) Travelling Light, Theme 2 of *T172 Working with our environment: technology for a sustainable future,* Milton Keynes, The Open University.

Potter, S., Enoch, M., and Fergusson, M. (2001) *Fuel taxes and beyond: UK transport and climate change*, London, World Wide Fund and Transport 2000.

Potter, S. (2003) Transport Energy and Emissions: Urban Public Transport, Chapter in Hensher, D. and Button, K. (eds) *Handbook in Transport 4: Transport and the Environment,* Pergamon/Elsevier.

Roy, R., Potter, S. and Yarrow, K. (2002) *Towards sustainable higher education: phase 1 final report*, Design Innovation Group, Milton Keynes, The Open University.

Royal Commission on Environmental Pollution (1994) *Transport and the Environment*, London HMSO, 1994.

Teufel, D. *et al* (1993) *Oko-Billanzen von Fahrzeugen*, Heidelberg, Umwelt und Prognose Institut.

Ubbels, B., Rietveld, P. and Peeters, P. (2002) Environmental effects of a kilometre charge in road transport: an investigation for the Netherlands, *Transportation Research Part D*, vol. 7, no. 4, pp. 255–64.

Watson, R. T. and the Core Writing Team (eds) (2001) *Climate Change 2001: Synthesis Report,* A contribution of Working Groups I, II and III to the Third Assessment Report of the Intergovernmental Panel on Climate Change (IPCC), Cambridge University Press.

Chapter 2

Sustainable Road Transport Techniques

by Ben Lane and James Warren

Figure 2.1 Congestion on a London road in 1919

Figure 2.2 Aerial view of a modern congested motorway – A Californian freeway, USA

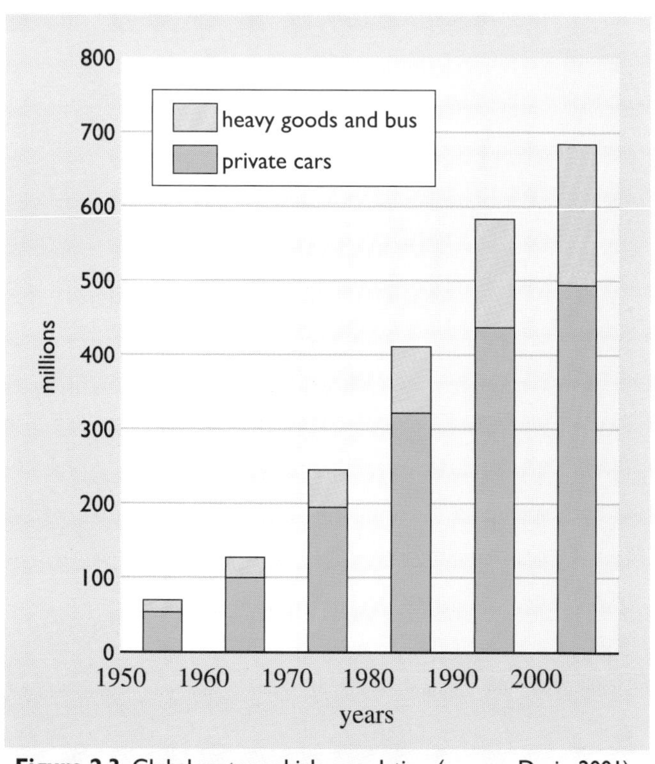

Figure 2.3 Global motor vehicle population (source: Davis, 2001)

2.1 Introduction

There is no doubt that the invention of the internal combustion engine, along with the extraction of petrochemicals to produce refined motor fuel, has significantly shaped society and the natural environment over the last one hundred years. In particular, the advent of motorised road transport, almost totally dependent on fuels derived from crude oil and the internal combustion engine, has transformed most aspects of modern life.

Book 1, Chapter 8 described the development of the early automotive industry. Once the technical obstacles had been overcome in the design of the 'Silent Otto' engine, it was the development of the moving production line by Ford in 1913 that first made the motorcar widely available. Within ten years, Ford were selling over a million cars per year, and in some parts of the United States, car ownership reached one person in three, a ratio only reached in the UK in the 1970s. Box 2.1 summarises some of the key dates that chart the early development of the motor car.

With the additional development of small diesel power units, and the discovery of large reserves of crude oil in many regions of the world, the automotive industry expanded rapidly throughout the twentieth century. During the last 50 years alone, the global vehicle population has increased by an order of magnitude to almost 700 million vehicles (see Figure 2.3). A similar increase has occurred in the UK, where the number of registered road vehicles in use is almost 30 million (DTLR 2001a). If the current rate of growth continues (at around 2% per annum), the global vehicle population could exceed 1 billion by the year 2020. There is, therefore, every likelihood that the motor vehicle will continue to significantly affect all aspects of life in the modern world.

BOX 2.1 Key events in the industrialisation of the motor car

1859 Accidental discovery of oil (whilst searching for water in Pennsylvania, USA)

1860 Etienne Lenoir patents the spark ignition engine

1876 August Otto produces first commercial 4-stroke engine

1892 Rudolf Diesel patents the compression ignition engine

1908 First Ford 'Model T' sold in the US

1913 Mass production of the Ford 'Model T' on the first modern production line

1924 MAN produces 5-litre diesel engine for road vehicle use

1925 Automotive sector becomes the largest industry in the US

1940 Over 200 cars per 1000 persons in the US

1960 Global car population exceeds 100 million vehicles

1974 Clean Air Act passed in United States

1992 Auto-Oil programme leads to first European vehicle emission standards (Euro I)

2000 Global car population surpasses 650 million vehicles

Today, the transport and petroleum sectors have grown to a point where air and surface transport accounts for almost 60% of global oil consumption and around a quarter of total energy consumption (IEA, 2002). The most common fuels for use by *road* transport are petrol and diesel, which are derived almost totally from crude oil. Both petrol and diesel require dedicated engine technology to convert the energy of these fuels into motive power. This chapter, therefore, begins by discussing the difference between the petrol and diesel internal combustion engines (ICEs).

2.2 Petrol and diesel engines

As discussed in Section 8.2 of Book 1, the petrol fuelled *spark-ignition* or 'Otto' engine (named after its inventor) is characterised by the use of a spark plug to initiate the combustion process. The engine utilises a four-stroke cycle, comprising the *induction, compression, power*, and the *exhaust* strokes. The four-stroke cycle is shown in Figure 8.2 in Book 1, Chapter 8.

Starting with the induction stroke, a small amount of fuel and air are drawn into the cylinder cavity. Whereas, older cars utilise a carburettor to mix the air and fuel to the correct ratio, modern vehicles tend to be equipped with fuel injectors where the air intake is via a high precision valve that sprays small amounts of petrol into the cylinder. This is usually done under the control of an electronic control unit (ECU). By using an on-board computer, the fuel injectors can vary the amount of petrol and air in order to achieve the lowest possible fuel consumption, or highest power output, depending on the engine load and accelerator position.

During the compression stroke, the petrol-air mixture is compressed into a small volume, usually to about a ninth of the original cylinder volume. In technical terms, the petrol engine is said to have a *compression ratio* of 9:1. (Typically, the air to fuel ratio is around 15:1 when measured *gravimetrically*.) The increase in pressure creates a highly explosive mixture, which is ignited by an electrical spark (from the spark plug). The gases burn very quickly causing a rapid expansion and a release of energy, which pushes the piston towards the connecting rod and crankshaft. This is the power or combustion stroke. Finally, the burned gases, which ultimately make up part of the exhaust, are flushed out of the cylinder during the exhaust stroke via the exhaust gas port or valve. The cycle then starts over again with another induction stroke.

The diesel engine works using the same four-stroke cycle as the petrol engine, but with two major differences involving the air-fuel mixture and injection systems. In the diesel engine, only the air is compressed in the cylinder instead of an air-fuel mixture, and at the end of the compression stroke, the fuel is directly injected into the combustion chamber by a fuel injection pump. Typical compression ratios of 20:1 are used, which is sufficient to raise the air temperature to over 400 ºC. Once the diesel fuel is injected into the cylinder, it immediately vaporises and spontaneously ignites. This combustion process produces a mixture of hot gases that then drive the piston. Diesel combustion is more explosive than for petrol. This leads to the characteristic diesel engine sound and explains why diesel engines are noisier and vibrate more than their petrol counterparts.

Modern diesel engines tend to use *direct injection* fuel delivery systems as they can be closely controlled by the use of computerised engine management systems. However, there are still many *indirect injection* diesel engines available, which work in the same way, but instead of injecting fuel directly into the combustion chamber, the fuel is injected into a pre-chamber before entering the cylinder. This allows for increased swirling (or mixing) that improves the combustion process. Older diesels also tend to be equipped with glow plugs, which heat the compressed air during a cold start by the use of an electrically-heated wire.

Figure 2.4 The Volvo 740 GLE injection engine

As the popularity of diesel engines has increased, several varieties of injection methods have been developed, including *common-rail* injectors, and *electronic unit injectors*. In the case of common rail, a single 'rail' or pipe is held at constant high pressure over the cylinders and a central control unit allows each injector to inject fuel electronically. Most current systems use a pressure of about 1350–1500 bar[1] to create two distinct fuel pulses: a pilot injection and the main (combustion) injection. The pilot injection helps seed the combustion process and can also be tuned to reduce engine noise. The next generation of engines will raise this pressure to at least 1800 bar along with multiple injections in order to lower emissions, engine noise and increase fuel efficiency without loss of overall engine power output. Electronic unit injectors are highly compact injectors that incorporate the fuel injection pump, the injector and the solenoid valve into a single unit. These injectors are mounted on top of each cylinder head and can deliver up to 2050 bar of pressure resulting in very efficient fuel use. Further advances in fuel injection technology are expected over the next few years due to increasing demands on the engine technology to be cleaner, quieter and more fuel-efficient.

[1] Atmospheric pressure at sea-level is approximately 1 bar (or 0.1 MPa in SI units).

In general, the fuel efficiency of a diesel engine is higher than for a petrol engine. This is primarily due to the fact that the combustion temperature (and pressure) within a diesel engine is higher than in a petrol power unit. This increases the engine's efficiently according to Carnot's equation for a perfect heat engine (see Book 1, Section 6.3). Other reasons are that diesel engines have less heat rejection (i.e. they cool more slowly) than spark ignition units and that diesel fuel has a slightly higher energy content than petrol on a volumetric basis (see Table 2.1). These thermodynamic differences also lead to different exhaust emission profiles for petrol and diesel engined vehicles.

In a diesel engine about 32% of the heat energy is delivered to the crankshaft, whereas in a petrol car only about 24% becomes delivered work. As this kinetic energy is delivered to the wheel via the mechanical *drive-train*, energy is 'lost' due to friction between the transmission components and due to aerodynamic drag. As a result, only about 24% of diesel fuel's energy ends up being used for moving the car. With petrol this is only 18%. Clearly, the actual values found vary enormously with the vehicle type and with the driving conditions (e.g. urban versus motorway driving). These figures could be considered relatively low, given the effort and cost associated with obtaining the fuel and manufacturing the vehicle in the first place. If we consider how much of the fuel's energy actually is used to move the payload, the situation is even worse. Taking into account the vehicle's mass, only around 1–2% of the fuel's energy is utilised to move the driver, passenger or freight.

Diesel's higher fuel economy (as compared to petrol) has been one of the reasons why there is an increasing demand (within Europe) for diesel cars. Other reasons include the facts that diesel engines provide increased low-end torque (more power at low engine revolutions) and higher peak engine power ratings. There is also a perception that diesel units are more durable than petrol power units. This *dieselisation* has seen an increase in the proportion of diesel cars in Europe from a base of 16% in 1985 to over a third in 2003. The proportion is predicted to continue to increase to 37% in 2004 (Standard and Poor, 1998). This has important implications, not only on energy consumption and fuel distribution, but also on local air quality composition (see the following section on Vehicle Emissions).

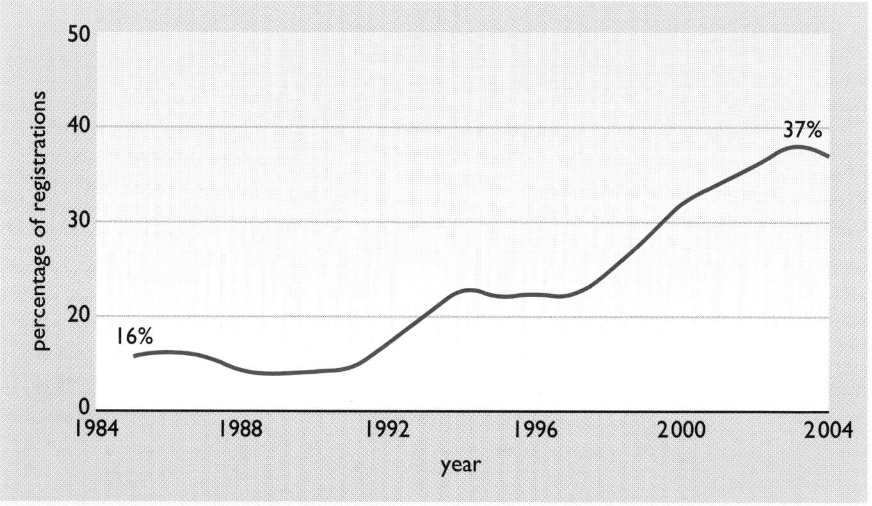

Figure 2.5 Dieselisation of the EU light-duty fleet (source: Automotive News Europe, 2001)

2.3 Petrol and diesel fuels

Petrol (known as gasoline or 'gas' in the US) and diesel are mixtures of liquid hydrocarbons refined from crude petroleum. The production of these fuels involves the extraction of crude oil, separation from other fluids, transport to refineries, processing (fractional distillation), transport to regional storage locations and distribution to retail or fleet refuelling stations. Each fuel must be carefully blended, either to control petrol's volatility and anti-knock performance (*octane* number) or diesel's ignition quality (*cetane* number).

Diesel fuel's main hydrocarbon chain is $C_{14}H_{30}$, and petrol has a main component chain length of C_9H_{20}. More energy is required to 'crack' crude petroleum to produce shorter chains of hydrocarbons. This explains why diesel needs less energy to refine than petrol (only about half as much) and why petrol has a lower *viscosity*. Table 2.1 compares typical properties of the two fuels.

Table 2.1 Properties of Petrol and Diesel Fuels

Fuel Property (units)	Petrol	Diesel
Hydrocarbon chain length	C_4 to C_{12}	C_3 to C_{25}
Carbon content by mass (%)	85–88%	84–87%
Fuel density (kg litre^{-1})	0.75	0.85
Lower heating value (MJ kg^{-1})	43.9	42.9
Lower heating value (MJ litre^{-1})	32.4	35.7

Sources: DTI, 2000 and AFDC, 2002

Internationally, there has been a trend to introduce *cleaner* conventional fuels through the removal of lead, sulphur and other additives and impurities. For example, whereas lead was originally added as an octane rating improver in 1923, unleaded petrol was first introduced in 1986 in the United States, with many countries following thereafter (Kitman, 2000). Indeed, leaded fuels were banned across the EU in 2000. European fuel specifications have also led to reduced sulphur and polyaromatic content. These include Ultra Low Sulphur Diesel (ULSD) (which also retails under name of *CityDiesel* in the UK) and Ultra Low Sulphur Petrol (ULSP). Current EU legislation allows the use of fuels with sulphur content up to 500 ppm. From 2005, all petrol and diesel fuels sold in the EU will have to qualify as ULSD or ULSP, with a maximum sulphur content of 50 ppm.

Some of the first countries to introduce ultra low sulphur fuels were Sweden in 1991 and Finland in 1993. The Swedish and Finnish governments supported the introduction of these fuels through the use of differential fuel duties. In other words, the cleaner fuels were charged less tax than standard fuels as an incentive to consumers to buy the cleaner grades. To support the introduction of Ultra Low Sulphur Diesel and Petrol in the UK, the British government also adopted this approach. Since 2000, all diesel sold in the UK has conformed to ULSD specification (five years ahead of EU legislation). ULSP is also being increasingly used in the UK and is likely to make up all petrol fuel in the very near future.

Figure 2.6 The range of fuels available at a typical petrol station

The main motivation for introducing reformulated fuels has been to reduce vehicle emissions. The reduction of sulphur in fuel significantly increases the longevity and efficiency of emission control systems (see Box 2.3) and reduces particulate emissions from diesel vehicles. However, removal of sulphur (and other impurities) is associated with an increase in *production emissions* and processing costs. These production impacts must therefore be taken into account in assessing the merits of a reformulated fuel (see Box 2.4).

2.4 **Petrol and diesel vehicle emissions**

During the firing of the first diesel engine in 1893, the inventor himself noted that '...black, sooty clouds came from the exhaust pipe in all of these tests' (Monaghan, 1998). Rudolph Diesel would have to work for another four years before this problem was addressed and, even today, the issue of pollution from the use of the combustion engine remains.

Conventional road transport leads to environmental pollution as a result of physical and chemical processes which occur during vehicle and fuel manufacture, production, use, recycling and disposal. As a rule, the energy consumed during a vehicle's manufacture is small in comparison to its energy use during its lifetime (Teufel, 1993; Mildenberger and Khare, 2000). Therefore, this section focuses on the emissions associated with vehicle use, which includes the impacts of fuel production and use (see Box 2.4). (There are also environmental impacts associated with road construction, road maintenance and the development of the transport and fuel infrastructure required by a road based transport system. However, these are not discussed in this text.)

As was discussed in Book 1, Section 8.4, within an internal combustion engine (in use), chemical processes take place between the hydrocarbons (HCs) of the fossil fuel, the fuel additives and the gases that naturally occur in the atmosphere (predominantly oxygen and nitrogen). The processes include complete and partial oxidation of the fuel, which produces carbon dioxide (CO_2), water (H_2O) and carbon monoxide (CO). Nitrogen from the air is also oxidised to nitrogen oxides (NO_x). Partially burnt and unburned fuel is present in the exhaust gases and forms a complex cocktail of volatile organic compounds (VOCs) together with small particles of matter ('particulates' or PMs), which are known to be prevalent in diesel fumes. *Tropospheric* or low-level ozone (O_3) is produced by the chemical action of sunlight on the VOCs, and subsequent reaction of the products with oxygen in the air. And in those countries that still permit the use of 'leaded' petrol, Lead (Pb) is also emitted with the exhaust gases. Box 2.2 provides a summary of the environmental effects of these emissions.

Petrol and diesel engines differ in their relative emissions performance with petrol vehicles emitting fewer NO_x and particulate emissions, and diesel vehicles producing less carbon dioxide per kilometre. As NO_x production is predominantly associated with reaction temperature, the relatively high ignition temperatures attained during combustion can explain diesel's higher NO_x. Diesel's lower CO_2 emissions are due to the engine's higher fuel economy as compared to petrol. Particulates up to 10 microns in size (termed PM_{10}) are also higher for diesels, although recent

BOX 2.2 Environmental effects of emissions associated with road transport

Carbon Monoxide	During respiration, carbon monoxide combines with haemoglobin in the blood, which hinders the body's ability to take up oxygen. This can cause and aggravate respiratory and heart disease.
Nitrogen Oxides	Responsible for acid deposition via the formation of nitric acid. Di-nitrogen oxide (N_2O; also known as Nitrous Oxide) contributes to global warming, and nitrogen dioxide (NO_2) is toxic to humans.
Particulates	Responsible for respiratory problems and thought to be a carcinogen. According to the World Health Organisation, there are no concentrations of airborne particulate matter (of size PM_{15} or less) that are not hazardous to human health.
Volatile Organic Compounds	Benzene and 1,3 butadiene are both carcinogens and are easily inhaled due to the volatile nature of these compounds. Other chemicals in this category are responsible for the production of ground level ozone, which is toxic in low concentrations. Also methane, released during the extraction of oil and during the combustion of petroleum products, is an important greenhouse gas.
Carbon Dioxide	The main environmental effect is as a greenhouse gas. In 2001, the International Panel on Climate Change (IPPC) concluded that doubling the amount of CO_2 in the atmosphere would be likely to increase the Earth's surface temperature by between 1.4 and 5.8 °C.
Tropospheric Ozone	In the stratosphere, ozone absorbs ultraviolet light, therefore reducing the number of harmful rays reaching living organisms on the Earth's surface. However, at ground level, ozone is toxic and responsible for aggravating respiratory problems in humans and reducing crop yields.
Lead	Lead is known to affect the mental development of young children and is toxic in small quantities. Originally introduced into petrol to improve its octane rating.

research suggests that petrol may produce more particulates in the $PM_{2.5}$ range.

Figure 2.7 compares petrol and diesel emissions from a typical small car with an engine size in the 1.4 to 1.7 litre range. Note the relative levels of CO, NO_x, PM_{10} and CO_2. To some extent, the emission profiles of petrol and diesel illustrate the general tendency for different conventional technologies to 'trade-off' one (or a group) of emissions against another. In this case local pollutants (NO_x, PM) are traded-off against global ones (CO_2). This inability of the internal combustion engine (ICE) to significantly reduce

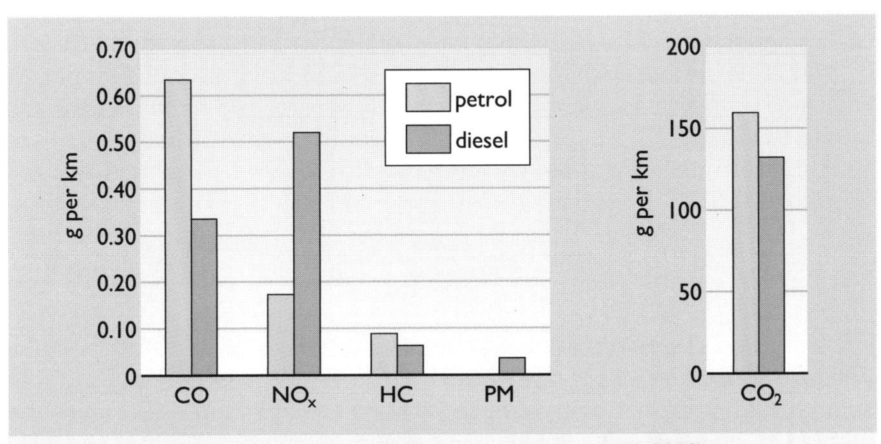

Figure 2.7 Vehicle emissions for a typical small car (source: DTI, 2000)

all emissions simultaneously *may* be an indicator that the technology is reaching the final stages of its development. However, only time will tell whether the ICE will be superseded by an alternative energy conversion device (see Sections 2.6 to 2.10).

During the last 30 years, several technological advances have significantly reduced the emissions from road vehicles. One of the most important emission control technologies has been the introduction of *three-way catalytic converter* (see Box 2.3 and refer back to Book 1, Section 8.4). These were first used in the United States in the 1970s so that vehicles would conform to the Clean Air Act, one of the first regulations that limited pollution from mobile (and stationary) sources. As a *technical fix*, these catalyst systems have done much to improve air quality over the years in the US, Japan and Europe.

BOX 2.3 **Catalytic Converters**

Catalytic converters are an important type of 'end of pipe' technology that reduces emissions of CO, NO_x and unburned hydrocarbons from the exhaust of petrol engine vehicles (and are hence known as 'three-way' catalysts). 'Cats' use a mixture of platinum, palladium and rhodium metals as their active components, which, in the presence of air, catalyses these emissions to less harmful gases. The catalysts are applied to a high surface area support structure (within the exhaust pipe) through which the exhaust gases are made to flow. The units are protected in a steel or metal canister, located within the vehicle's exhaust pipe.

Most systems have to meet stringent government lifetime requirements such as 100 000 miles, or 10 years of useful working life. Converters do have some inherent drawbacks. They are relatively ineffective before the 'light-off' temperature is reached, which means that they are inactive during short trips. Also, they tend to slightly increase fuel consumption (and hence carbon dioxide emissions). The precious metals in the converters can also be poisoned by certain fuel components such as lead and sulphur, which is why the use of catalysts has been dependent on the availability of lead-free and low sulphur fuels.

As in the US and Japan, European legislation continues to be tightened for *vehicle emissions* (see Box 2.4) and has been highly successful in reducing some of the pollutants associated with road transport. In Europe, the 'Auto-Oil' programme (a tripartite project involving the European Commission, oil and motor industries) has led to the introduction mandatory limits for what are termed the '*regulated emissions'*. These are carbon monoxide (CO), nitrogen oxides (NO_x), hydrocarbons (HC) and particulate matter less than 10 microns in size (PM_{10}). In particular, key legislation (for passenger cars) was introduced in 1992 (known as Euro I), in 1997 (Euro II), in 2001 (Euro III) and new legislation is due for 2006 (Euro IV) (see Table 2.2). Similar European limits have been introduced for heavy-duty vehicles (specified in terms of grams per kWh of engine output).

It is interesting to note that, although transport is responsible for around a fifth of CO_2 emissions in the UK, there is no current legislation that limits the amount of carbon dioxide produced per km for road vehicles. However, the European Commission's *target* is to reduce emissions of CO_2 from new cars sold in the EU to an average of 140 g km^{-1} by 2008 and 120 g km^{-1} by 2012.

Table 2.2 Existing and future European emissions limits for passenger cars

Emissions Limits	**Petrol (g km^{-1})**				**Diesel (g km^{-1})**			
	CO	HC	NO_x	$HC+NO_x$	CO	NO_x	$HC+NO_x$	PM
Euro II (1997)	2.20			0.50	1.00		0.70	0.080
Euro III (2001)	2.30	0.2	0.15		0.64	0.50	0.56	0.050
Euro IV (2006)	1.00	0.1	0.08		0.50	0.25	0.30	0.025

Source: DTI, 2000

This would represent a cut of around 25% of the current average. To achieve this aim, the Commission has reached a formal (though voluntary) agreement with ACEA (the European car manufacturers' representative organisation) to implement the required technologies to reduce carbon emissions.

For petrol and diesel vehicles, carbon emissions are closely correlated with fuel use. Therefore, the trends in carbon dioxide emissions are similar to those for fuel economy. Given the facts that engine *designs* are becoming more efficient, and cars are becoming lighter and more aerodynamic, one might think that the overall fuel economy is improving and CO_2 emissions, on average, are decreasing. However, this is not the entire story. Many engines are in fact becoming more powerful, especially in the lower end of the torque range, and at higher cruising speeds. This trend is partly driven by consumer demand for more powerful cars, and for more extra features within the vehicle. Equipment such as air conditioning, heated seats, electric windows, auto-defrosting, and on-board navigation all require energy to operate. As of 1999, the average new (petrol two-wheel drive) car fuel consumption was 7.3 litres per 100 km (or 38.5 mpg) which is equivalent to around 180 g of CO_2 per km, so there is a long way to go to reach the 2012 CO_2 emissions target. Even the most efficient diesel and petrol cars today average 58–64 mpg, equivalent to 115–120 g of CO_2 per km, which is an excellent start to reducing energy consumption, but a long way from ensuring that all cars meet the target.

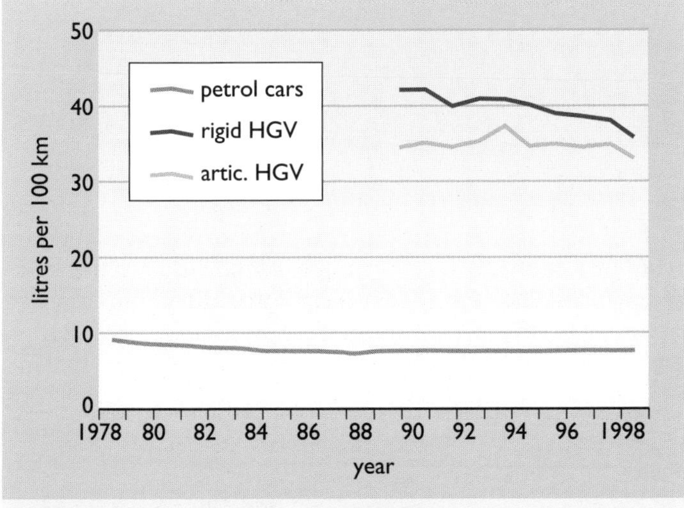

Figure 2.8 Vehicle fuel economy 1978–99

BOX 2.4 Life-cycle analyses of vehicle emissions

The *life-cycle* analysis of road transport emissions is an established methodology that has been used by many researchers to compare the environmental impact of different road vehicle fuels and technologies (ETSU, 1996; Hart and Bauen, 1998b; DTI, 2000). A full analysis of road transport emissions needs to account for both *vehicle emissions* generated during vehicle use ('tank-to-wheel') and *production emissions* generated during fuel production, processing and distribution ('well-to-tank'). The total emissions due to fuel production and vehicle use are termed *life-cycle emissions* (also known as 'well-to-wheel').

Emissions (and other environmental impacts) are also associated with *vehicle manufacture*. Though they are not insignificant, they are not usually included in a comparison of fuels or vehicle technologies, unless it is thought that the difference in vehicle production methods is very different from conventional manufacturing processes. This can be the case where radically new technologies are being considered (e.g. change from ICE to electric vehicle technology). However, emissions associated with vehicle manufacture are not included in this Chapter as the energy consumed during vehicle manufacture is small in comparison to lifetime energy use (Teufel, 1993; Mildenberger and Khare, 2000).

Note that vehicle emissions are specified in grams per kilometre for light-duty vehicles and grams per kWh (engine output) for heavy-duty vehicles. Production emissions are specified in grams per unit of energy delivered (usually GJ).

2.5 Cleaner conventional vehicle technologies

Vehicle emission legislation has been one of the strongest factors forcing car manufacturers and their suppliers to develop less polluting engines. Technology improvements to date include more efficient engine designs, new tail-pipe emission control and electronic management systems, and improved sensing devices to monitor the state of the engine and exhaust.

For some time, some manufacturers have been developing the next generation of petrol power units, which include Gasoline Direct Injection (GDI) engines. These were first developed by Mitsubishi who claim that the technology offers up to 20% fewer carbon emissions and a similar improvement in fuel efficiency. A GDI engine works like a normal petrol power unit, except that the petrol is sprayed directly into the combustion cylinder (there is no pre-mixing stage as in a typical petrol engine). This results in a cleaner burn and an increase in power. The main obstacle to this technology is the sulphur content of petrol, which, even at the levels found in ULSP (50 ppm), hinders the catalysts that are required to control the NO_x emissions (see comments on the lean burn engine in Book 1, Section 8.4).

In addition to the three-way catalytic converter, much work has been conducted to develop new exhaust emission control systems for petrol and diesel engines. One such device is a *particulate filter* used for the Peugeot 607 diesel car. This is a complex system containing a filter to trap the soot, an additive that helps burn the trapped particles and a control system to monitor the soot level and initiate combustion of the particulates when required. For heavy-duty engines, emission control devices include oxidation ('one-way') catalysts, Exhaust Gas Recirculation (EGR) systems and continuously regenerating traps (CRTs). These devices are increasingly being fitted as standard and are proven to reduce particulates by up to

90%. It is widely accepted that these technologies will be required if the heavy-duty diesel sector is to be able to comply with Euro IV standards due in 2006 (DTI, 2000).

Table 2.3 Future likely improvements in vehicle ICE design

Technology type	Diesel	Petrol
Increasing Exhaust Gas Recirculation (EGR)	✓	✓
Higher injection pressures	✓	
Particulate trapping	✓	
Variable pressure turbo-charging		✓
Cylinder deactivation		✓
Heated catalysts		✓
Engine downsizing (with electronically assisted turbocharger)	✓	✓
Use of lightweight components (aluminium, plastics)	✓	✓
Automated manual transmissions	✓	✓
Integrated starter/alternator units	✓	✓

Table 2.3 shows some of the technology improvements that are likely to continue to be introduced for diesel and petrol ICE vehicles in the next decade. Although many of these advances will be effective at improving fuel use and reducing vehicle emissions, a recent technological breakthrough may prove to be even more productive (and cost-effective). This approach has returned to the drawing board to address one of the intrinsic incompatibilities of conventional engine design and vehicle use; namely that, for a conventional ICE vehicle, the maximum efficiency of the engine is only achieved when the engine's output is matched to the load (see Book 1, Section 8.4). This usually occurs at 2000–3000 engine revolutions per minute (rpm) at a moderate cruising speed of 100 kph on a level road. These 'perfect' engine conditions are rarely achieved during urban driving which includes low average speeds and stop-start traffic. As a result, the average engine efficiency falls far short of its design optimum (as do the fuel economy and vehicle emissions). To address this issue, some manufacturers are developing a new engine configuration which increases the time an ICE engine can operate close to its point of maximum efficiency; the *hybrid petrol-electric vehicle.*

The hybrid petrol-electric vehicle

The ICE vehicle can be combined with a battery electric traction system in what is called a *hybrid electric vehicle* (HEV). In a very real sense, a hybrid is part conventional ICE technology and part electric vehicle (see also following section for *battery electric vehicles*). The principle underlying all hybrid vehicles is that the use of an energy buffer (usually a *secondary-cell*, also known as a rechargeable battery) which enables the main power unit to be operated at close to maximum efficiency. When the engine loading is low, the excess output is stored as chemical energy (within the battery) for later use. When the loading is high, the main power unit and the battery work together to deliver the required power. The use of an on-board battery also enables the use of *regenerative braking* which recovers part of the energy usually 'lost' during braking, so reducing overall energy use. In this way, HEVs provide significantly improved fuel economy and reduced emissions.

There are various types of hybrid vehicles that are usually categorised as *parallel*, *series* or *split* hybrids. Each of these definitions refers to the configuration of the main and peak power units. Briefly, the three different systems can be summarised simply as:

- **Parallel** – the engine and electric battery/motor are both connected to the transmission, so that either the engine or the motor can provide power to the car's wheels.
- **Series** – the engine does not directly provide power to the car's wheels; instead the engine drives a generator, which can power the motors that run the wheels or charge the batteries.
- **Split** – the engine drives one axle whilst the electric motor drives the other. There is no connection between the mechanical and electric drive-trains.

A technical break-through in hybrid vehicles occurred during the 1990s, when several motor manufacturers developed HEVs to (or almost to) production stage. The first commercially available *petrol-electric hybrid*, the Toyota Prius, was launched in Japan in 1997 and in Europe in 2000 (see Book 1, Chapter 1, Figure 1.51). Hybrid passenger cars have also been launched by Honda (the Insight and Civic), and are being developed by other manufacturers including Nissan, Audi, Renault, Peugeot and Volkswagen. Hybrid bus and truck vehicle projects are also underway in many European cities.

The Prius' hybrid system incorporates a 53 kW 1.5-litre petrol engine rated at 4500 rpm and a 33 kW electric motor. The Prius can be categorised as either a series or a parallel hybrid due to the unique nature of an electronically controlled 'power splitter'. This device allows the hybrid system to direct power from the conventional engine to either the wheels or to the generator. The generator in turn can drive a motor (to power the wheels) or to charge the battery. The battery is used to drive the motor when the engine is off or needs extra power.

As of 2003, compared to conventional vehicles, global production of the Prius is relatively low at around 1500–2000 cars per month. However, already the car is outselling other hybrid models and other types of battery electric vehicle (see next section). Given this initial, though modest, success, many of its innovative features are likely to become standard in other hybrid vehicles over time.

Under test, the Prius petrol-hybrid shows that all regulated emissions are significantly reduced as compared to an equivalent petrol ICE vehicle. In 2000, the car already complied with the Euro IV standard (six years ahead of EU legislation). Research in the US also shows that hybrids are more fuel-efficient than a conventional diesel vehicle (and emit less CO_2) by 20–30% (Cuddy and Wipke, 1998). This is confirmed by the Prius, which achieves a fuel economy of 4.9 litres/100 km (58 mpg) on a standard European drive cycle. This represents a fuel economy improvement of around 30% as compared to a petrol equivalent. Given that the associated CO_2 emissions are around 114 g CO_2/km, hybrids may provide the auto

BOX 2.5 Innovative features of the Toyota Prius petrol-electric hybrid

- **Main power unit** – engine is sized (53 kW) for the average load instead of the peak load, which is only required 1% of driving time.
- **Peak power unit** – a battery is used to drive a 33 kW motor when extra power is required or when the main engine is in 'off' mode.
- **Power splitter** – directs main engine power to either the wheels or the generator, which in turn can drive a motor (to power the wheels) or charge the battery.
- **Regenerative braking** – during braking, the motor acts as a generator, providing energy to recharge the battery; most systems can recover around 20% of the energy normally 'lost' during braking.
- **Pure electric operation** – the main engine switches off when not required, resulting in a much lower fuel use.
- **Reduced emissions** – efficient fuel use together with an efficient, high-density catalyst reduce the regulated emissions by up to 90%. Fuel use and carbon dioxide emissions are reduced by around 30%.

industry with a technology that can deliver the EU/ACEA targets for vehicle carbon emissions (see Section 2.4).

As with most new technologies, the benefits of hybrids come at a price. In 2001, the Prius retailed in the UK at £16 430 compared with an equivalent Toyota petrol car (the Carisma) which cost around £13 000. However, the Prius does reduce fuel costs that offset its higher capital cost, as the reduction of fuel use by 30% translates directly to a fuel cost saving of the same amount. Government grants are also available (£1000 per new hybrid car) to assist purchasers of this new technology (see http://www.powershift.org.uk/ [accessed 12 June 3003]).

As of 2003, hybrid-electric vehicles are still at an early stage of development and a dominant system design (if one exists) has yet to emerge. Thus, equal numbers of series and parallel systems are under development. Generally, there seems to be a consensus that parallel hybrids will dominate the market, initially, followed by series designs. It is difficult to predict the future success of hybrid technology until the first commercially available vehicles have been used in real driving conditions over an extended period of time. However, early experience has been very encouraging and hybrids possess great potential to become the standard automotive technology during the coming decade.

2.6 Alternative vehicle fuels and engines

As we have seen in previous sections, conventional vehicle systems are based on petrol and diesel fuels and on the internal combustion engine. During the past century, these fuels and technologies have become highly developed and are supported by global industries that have made vehicles affordable to most people in the modern world. There have also been improvements to engine efficiency during that time, which has allowed the addition of on-board devices that improve driver and passenger comfort and safety. From an environmental perspective, the regulated emissions have also been reduced by over an order of magnitude (on a per km basis).

However, limits to the development of the conventional ICE vehicle have become apparent. Overall, car fuel economy has not improved significantly over the past few decades (see Figure 2.8) and average engine efficiencies remains at around 15–20%. Worse, as already mentioned, due to the relatively high mass of vehicles, only a few percent of petrol or diesel's energy is utilised to actually move the driver or payload. The motorcar is also totally dependent on the supply of crude oil, which makes the use of motor transport a highly political issue. Lastly, the transport sector is a significant contributor to total greenhouse gas emissions (around 20% in the UK), and even though regulated emissions are reducing *per vehicle*, NO_x and particulates remain a problem in many cities due to the increasing number of vehicles on the road.

For these reasons, governments, together with the fuel and automotive industries, have been attempting to develop **'alternative' vehicle fuels**, which would reduce dependence on oil and/or lessen road transport's environmental impact. This process has started with the introduction of ultra low sulphur diesel and petrol. But truly alternative fuels may provide further benefits in the longer term including air quality and climate change benefits. The advantage of alternative fuels is that they can be used in 'conventional' internal combustion engine vehicles. Their use also improves the use of advanced after-treatment systems that can further reduce vehicle emissions.

Cleaner fuels that provide tangible emissions benefits include **natural gas, liquefied petroleum gas, bio-fuels** and **hydrogen.** These are already being used for transport applications and have been shown to reduce vehicle and *life-cycle emissions* on a per km basis (see Box 2.4). In principle, they can be used in most ICE vehicles with relatively minor modification. Indeed, as was introduced in Book 1, Section 14.5, BMW have developed a series of hydrogen ICE prototype cars, the latest of which includes the hydrogen powered Mini-Cooper.

A second (and more radical strategy) is to develop **'alternative' vehicle technologies.** These involve the use of totally new energy conversion systems that partially or completely replace the internal combustion engine (ICE). In fact, this process has already begun with the introduction of the **petrol-electric hybrid vehicle** (see last section), which, as the name suggests, combines the advantages of an ICE with that an electric drive-train. Indeed, most of the alternative vehicle technologies under development employ electric (as opposed to mechanical) drive-trains. These include **battery electric vehicles**, which are particularly suited for urban and short-range use and the **fuel cell electric vehicle,** which is (usually) fuelled by hydrogen and which has been considered by many in the motor industry to offer great potential as a road vehicle technology. As mentioned in Book1, Chapter 14, fuel cells have been used for space exploration since the 1960s. Indeed, the more general *hydrogen economy* has been considered by some analysts to be 'inevitable', providing a means of long-term storage for renewable energy (Serfas *et al.*, 1991). Though hydrogen can be used within existing ICEs, *hydrogen fuel-cell vehicles* would radically change patterns of transport energy use and environmental impact.

Table 2.4 shows the alternative fuels and vehicle technologies that will be discussed in the following sections. The options considered by no means form an exhaustive list, but they do represent the alternatives considered by most analysts to have the potential to be commercially viable within Europe by 2010.

Table 2.4 Alternative vehicle fuels and technologies discussed in Chapter 2

Alternative vehicle fuels (utilising mechanical drive-trains)	Alternative vehicle technologies (utilising electric drive-trains)
Compressed Natural Gas (CNG)	Hybrid-Electric Vehicles (HEV) – see Section 2.5
Liquefied Petroleum Gas (LPG)	Battery Electric Vehicles (BEV)
Bio-fuels (bioethanol, biomethanol and biodiesel)	Fuel Cell Electric Vehicles (FCV)
Hydrogen	

2.7 Natural gas and liquefied petroleum gas

Compressed Natural Gas (CNG) and Liquefied Petroleum Gas (LPG) are mixtures of low-boiling temperature hydrocarbons. The main constituent of natural gas is methane (CH_4) with smaller amounts of propane (C_3H_8) and other hydrocarbon gases. LPG is a mixture of propane (over 90% in UK) and butane (C_4H_{10}). Being relatively simple chemical compounds that mix easily with air, these gases enable a more complete combustion than do conventional liquid fuels, which can lead to a reduction in vehicle emissions. The gases also have high *octane ratings* that enable a high compression ratio to be used, so improving engine efficiency.

Table 2.5 Properties of Alternative Fuels

Fuel Property (units)	CNG	LPG	Methanol	Hydrogen
Chemical formula	CH_4	C_3H_8	CH_3OH	H_2
Carbon content by mass	75%	82%	37.5%	0%
Typical storage pressure (bar)	200	8	1	200
Fuel density (kg/litre)	n/a	0.51	0.80	n/a
Lower heating value (MJ kg^{-1})	47.6	46.4	19.9	120
Lower heating value (MJ $litre^{-1}$)	n/a	23.6	15.7	n/a

Note that fuel density and volumetric heating value cannot be specified for gaseous fuels. Sources: DTI, 2000 and AFDC, 2002

Natural gas vehicles (NGVs) were first introduced in Italy just before World War II, for use in light-duty commuter cars, and were supported through the use of government subsidies. The Argentinean government were also early promoters of the fuel, partly in response to severe air pollution problems in Buenos Aires, and partly to conserve their own supplies of oil for export to earn foreign currency (IEA, 1999). Currently, over 1.2 million NGVs are in use worldwide in over 40 countries with Argentina, Russia, Italy, Canada and the USA operating the largest fleets. Excluding Italy, Europe has over 8500 NGVs, serviced by 175 filling stations. The UK has over 750 NGVs, which are mainly operated by fleet operators in the private and public sectors.

In the Netherlands, LPG is already considered a 'conventional' motor fuel with most Dutch motorway filling stations supplying the fuel and around 6% of the light-duty vehicles using the fuel. Several major Dutch cities have public transport fleets operating on LPG and it is common for transport companies to buy buses for conversion to LPG. Worldwide, there are currently over 4 million LPG vehicles with 2 million in Europe alone; mainly in Italy, Holland, former Soviet Union, Japan, USA and Australia. In the UK, there are over 50 000 LPG vehicles on public roads, the majority being light-duty vehicles that have been converted to run on LPG fuel.

Vehicle technology

Most light-duty vehicles that operate on 'road gas' fuels are *bi-fuel* conversions. These utilise a traditional spark-ignition engine optimised for petrol, which is also able to run on LPG or natural gas. Whereas older conversions often had poor performance (the engine being optimised for petrol operation), recent conversions incorporate fuel-injection systems that have greatly improved engine response for both fuels. However, some drivers of bi-fuel vehicles continue to report some power loss when using gas. For this and other reasons, *dedicated* gas engines maximise the benefits that are offered by LPG and natural gas and can provide vehicle performance similar to conventional fuels. In many cases, improvements in engine performance are found for heavy-duty conversions to gas, including higher torque at low rpm and an extended engine life due to the cleaner fuel and reduced engine stress.

Compressed natural gas (CNG) is normally stored on-board a vehicle in a pressurised tank at around 200 bar. Cars are typically fitted with a single cylinder that contains 16 kg of gas, equivalent to the energy of 23 litres of petrol. For steel cylinders, which are most common, the combined tank-fuel weight is about four times heavier than for petrol/diesel. This increases fuel consumption and reduces the payload that can be carried. Therefore dedicated NGVs tend to be heavy-duty vehicles where the extra weight and volume of the gas tanks is less of an issue. LPG can be liquefied more easily than natural gas and is stored as a liquid under moderate pressure (at 4–12 bar). As LPG storage tanks pose less of a space problem than does CNG, LPG has become very popular within the light-duty sector in the UK. Uptake has also been promoted through the low cost of conversion, the ease of refuelling and the increasing availability of the fuel.

Though emissions benefits are offered by use of NG and LPG vehicles (*see below*), their use is associated with increased capital costs. For example, the additional costs for a NG storage cylinder (for a heavy-duty vehicle) can be as high as £10 000. Even for cars, CNG adds 10–15% to the cost of a vehicle. Conversion to LPG is less expensive for light-duty vehicles at around £800–£1500 for cars and vans and around £15 000–£25 000 for bus conversions. To offset higher capital costs, Government subsidies are available to assist with gas vehicle purchase or conversion. Under the PowerShift programme, the government provides a subsidy of up to 75% of the extra cost as compared to a petrol/diesel vehicle. For more information visit http://www.powershift.org.uk/ [accessed 12 June 3003].

Fuel supply and infrastructure

NG re-fuelling systems can either be 'fast-fill', using gas at 250 bar to refuel a vehicle within minutes, or 'slow-fill' which uses a compressor to 'trickle charge' a vehicle over several hours. LPG is dispensed as a liquid under moderate pressure in much the same time it takes to refuel a petrol or diesel vehicle. In the UK, while there are only around 20–30 CNG filling points, there are over 1100 LPG stations. This explains in part why LPG has become the more popular gaseous fuel for light-duty use. However, although public re-fuelling facilities can service more vehicles, depot-based refilling sites are playing an important role in the development of NG, LPG (and other) cleaner fuels. This is because fleets using alternative fuels can be more easily managed using centralised re-fuelling and support facilities.

Figure 2.9 Refuelling a bus powered by natural gas

The high capital cost of NG refuelling systems also acts as a barrier to the up-take of road gases. For example, in Southampton, £250 000 was required for a system to fast-fill a fleet of 16 buses. Infrastructure costs are less of a problem for LPG as the fuelling units operate at lower pressure than for natural gas, which again explains why the up-take for LPG has been initially greater than for CNG. However, the increased cost of gas vehicles and fuel infrastructure is partially offset by the relatively low *price* of gaseous fuels, which have benefited from advantageous fuel-duties set by national government. (The UK fuel duty on gaseous road fuels was cut from 21p/kg to 9p/kg over the period 1998–2001. The UK government has also pledged to freeze fuel duty on NG and LPG until 2004.) LPG forecourt prices are around 38 p/litre (about half price of petrol). Taking into account the low energy density of LPG, the *petrol-equivalent* price of LPG is around 49p/litre. The forecourt price of natural gas is around 61 p/kg, equivalent to 41p/litre petrol or 45p/litre diesel.

Environmental impact

For light-duty vehicles, with the exception of hydrocarbons, the *regulated* emissions are significantly reduced for gas-powered vehicles. Compared to petrol-car emissions, CO and NO_x are reduced by at least a third and particulates are virtually eliminated. Hydrocarbons are almost halved for LPG vehicles, whereas these emissions can be increased for some NGVs due to the presence of non-combusted methane in the exhaust gases (which has significant implications for greenhouse gas emissions).

For dedicated heavy-duty vehicles, as compared to diesel, the reductions are around two-thirds for both NO_x and particulates. Emission of hydrocarbons are reduced by well over 50% for LPG, though are significantly higher for heavy-duty NGVs. However, over 80% of these HC emissions are composed of methane that can be almost eliminated from exhaust gases by the use of dedicated catalyst systems.

Energy use per km is slightly increased for gas operation as compared to conventional fuels. However, due to the NG and LPG's low carbon content (see Table 2.5), vehicle CO_2 emissions (per km) are reduced. For light-duty vehicles, tests provide evidence of a 10%–15% reduction of life-cycle CO_2 emissions as compared to petrol operation and up to 5% reduction as compared to diesel. For heavy-duty vehicles, life-cycle CO_2 emissions are comparable to diesel operation.

In assessing the full impact on global emissions, it should be remembered that methane is an important greenhouse gas. Therefore, for NGVs, the methane emissions from the vehicle, refining and distribution processes must be accounted for in the calculation of the effect on global warming. The result is that, for heavy-duty NGVs, total lifecycle greenhouse gas emissions are comparable or *slightly increased* when compared to diesel operation. This situation will improve as more dedicated gas engines are brought on to the market with optimised catalysts.

BOX 2.6 Pros and cons of road fuel gases

Advantages:

- **Reduced emissions** – reduced NO_x, particulates and CO_2 (for cars).
- **Reduced fuel costs** – up to 25% lower fuel cost per km.
- **Reduced low noise levels and engine vibration** – noise reduction of 68 dB to 60 dB for heavy-goods vehicles (equivalent to a car) (for explanation of dB scale – see Book 2).
- **High fuel availability** – UK national gas grid is already in place and LPG available at over 1100 refuelling stations.

Disadvantages:

- **Higher capital costs** – 10%–15% higher vehicle costs for light-duty conversions and up to £25 000 additional costs for heavy-duty dedicated gas vehicles.
- **Poor NG refuelling infrastructure** – although NG is available through the national grid, very few filling station have been installed due to high equipment costs.
- **Reduced vehicle payload** – mass and volume of gas tank can reduce payload capacity of heavy-duty NGVs by up to 1 tonne.
- **Vehicle restrictions** – some restrictions in use of LPG in confined spaces (tunnels, car parks) within Europe.

BOX 2.7 LPG vehicles: Oxfordshire Mental Healthcare NHS Trust

Oxfordshire Mental Healthcare NHS Trust operates health services from various sites throughout the county. Small passenger vehicles meet most of their transport requirements. The Trust operates a regular minibus service acting as a non-stop shuttle between its sites. The bus service ferries members of staff to and from work and delivers cost and environmental benefits in reducing staff reliance on private car use.

This Transit minibus service travels around 44 000 miles a year using two drivers who provide the service five days a week. This equates to around 170 miles every working day. As well as passengers, the vehicle also carries pre-prepared meals and internal post. A desire to take action on environmental and congestion problems initially led the Trust to investigate the benefits of a clean fuel minibus service. Oxford City Council has adopted its own green transport policy that includes the use of clean fuel vehicles, and the Trust is keen to align itself with these local activities.

The Ford Transit minibus has a standard vehicle specification, and did not require any special modifications apart from the LPG conversion. One seat has been removed making way for a permanent storage area specifically designed to carry parcels and hot pre-cooked meals. The Transit minibus cost £14 000 with the LPG conversion costing an additional £1700. However, this was eased somewhat by a PowerShift grant of £1000. The LPG vehicle would be expected to reduce by half the carbon monoxide, hydrocarbons, and oxides of nitrogen emissions of a comparable petrol vehicle operating in the streets of Oxford.

The minibus has proven to be very suitable for its specific operation but the Trust is convinced it could be used for other duties if so required. The vehicle's performance has been satisfactory and meets the needs and requirements established in the Trust's Transport Plan. The transport manager also considers that: '*Considering the high mileage and punishing duty cycles we subject the vehicle to, it has proven very suitable for its use*'.

The Trust found that it was much cheaper to install LPG refuelling facilities at its main site, than to refuel at existing local alternatives. The LPG tank was supplied and installed on-site by Calor and no additional infrastructure was required. The vehicle is not refuelled at any other LPG station whereas petrol, when required, is supplied through local retail outlets. The LPG fuel cost is very competitive and there is an additional nominal rental fee for the gas storage tank. Significant cost savings over a conventional diesel vehicle have been achieved.

Overall vehicle reliability has been very good. Although the Trust does not operate a conventional minibus for comparison purposes, the reliability of the LPG minibus has not caused any concern. Spark plugs have required changing every 10 000 miles as opposed to the 20 000 envisaged. Basic LPG fuel system maintenance and repairs have been undertaken by a trained engineer working for the local ambulance service. When more awkward problems arise it can take up to two days to get an appropriate LPG engineer. Ford had been approached regarding the possibility of extending the vehicle's warranty cover beyond one year. They agreed to an extension but declined to cover any of the LPG fuel system components.

In addition to the LPG Transit minibus, the Trust also operates a LPG Transit van used for general maintenance duties. There are several vehicles operated by the Trust that will soon need to be replaced. LPG will be seriously investigated as an alternative to the current diesel options. According to Neil Godfrey, the Trust's transport manager, leasing companies have not, until recently, been interested in supplying CFVs [Cleaner Fuelled Vehicles] but: '*Now that mainstream automotive manufacturers are producing bi-fuel, as well as electric vehicles, things should change*'. The Trust remains firmly behind its decision to invest in CFVs. With more certainty about wider refuelling options, further expansion of the clean LPG fleet is likely.

Key facts

Featured vehicle	Bi-fuel Minibus
Conversion cost	£1700
PowerShift grant	£1000
Average monthly mileage	3000 miles
LPG fuel cost	22.5p per litre (bulk purchase)
Economy Approx.	20% less mpg of fuel compared to conventional vehicle
Emissions	Significant reductions in particles and oxides of nitrogen
Performance	Unchanged from petrol vehicle
Passenger numbers	11 people

Source: EST, 2001b

2.8 Biofuels

Liquid biofuels are produced by the *fermentation* of energy crops or the *esterification* of vegetable oils or animal fats. These fuels can reduce the transport sector's dependence on fossil fuels, and, in principle, their use can provide reductions in some regulated and greenhouse gas emissions on a life-cycle basis.

Ethanol (CH_3CH_2OH; also known as *ethyl* or *grain* alcohol) is a clear, colourless liquid and is the essential ingredient of all alcoholic drinks. It can be produced from virtually any fermentable source of sugar. Ethanol made from cellulosic biomass materials instead of traditional feedstocks is called **bioethanol**. The production method first uses enzyme *amylases* to convert the feedstock into fermentable sugars (dextrose). Yeast is then added to the *mash* to ferment the sugars to ethanol and carbon dioxide (CO_2).

Figure 2.10 A substantial proportion of vehicles in Brazil are fuelled by alcohol [gasohol] derived from sugar cane

Another alcohol fuel is methanol (CH_3OH; also known as *wood* alcohol), which is predominantly produced via steam reforming of natural gas to produce *syngas* (a mixture of carbon monoxide and hydrogen). This is then fed into another reactor vessel under high temperatures and pressures, where the gases are combined in the presence of a catalyst to produce methanol and water. Although over 80% of methanol is currently produced in this way, the ability to produce **biomethanol** from non-petroleum feedstocks (including biomass) is of interest for reducing reliance on fossil fuels.

Biodiesel is most commonly produced by the *esterification* of energy crops such as oil seed rape (OSR) or recycled vegetable oils (RVO). Animal oils can also be used (see Box 2.8). The oils are filtered and pre-processed to remove water and contaminants and are then mixed with an alcohol (usually methanol) and a catalyst. The oil molecules (tri-glycerides) are broken apart and reformed into fatty acid methyl esters and glycerol, which are then separated from each other and purified. Biodiesel from OSR is known as Rape Methyl Ester (RME). The fuel can be used (pure or as a blend) in place of mineral diesel without any modification in modern diesel powered vehicles. The production of RME also has two valuable by-products, glycerine, which is used in pharmaceuticals and cosmetics, and cattle cake made from the remaining plant material.

BOX 2.8 Chicken fat to power lorries

Starting in January [2003], Asda trucks of up to 40 tonnes will carry startling slogans saying 'This vehicle is powered by chicken fat' – the biggest boost yet for the legal use of recycled cooking oil on Britain's roads. Lorries making deliveries on Tyneside and in Yorkshire will be the first to try the fuel, which is currently available on three forecourts in Yorkshire. A further eight garages in the region are to take supplies from the growing number of biodiesel refiners, who were given a 20p-a-litre green tax concession by the chancellor, Gordon Brown, in July [2002].

Asda produces more than 50m litres of used cooking oil and 138 000 of waste frying fat every year from its canteens, restaurants and rotisseries. The gunge was a disposal headache rather than a potential money-earner until an unexpected phone call last spring. 'We were approached by a biodiesel firm, which cleans up waste cooking oil, adds a bit of methanol and sells it as a much cheaper alternative to diesel,' said Rachel Fellows of Asda yesterday. 'We were only too happy to do business with them. 'But then we thought: hang on, isn't there something we can do here for ourselves?'

Company trials of 'chip pan fuel' for Asda's cars and lorries were then intensified after the firm's innocent involvement last month in a moonshine operation at Llanelli in South Wales. A special 'frying squad' set up by Dyfed Powys police discovered that hundreds of drivers were running their cars on Asda's 'extra-value' cooking oil mixed with methanol at home, in a moonshine operation which dodged tax. The 32p-a-litre fuel supply – compared with 73p at forecourt diesel pumps – was cut off when Asda discovered its Llanelli branch was selling vastly more oil than anywhere else in the country. Rationing was imposed and the police frying squad – whose tactics included sniffing out the chip-shop smell of bootleg cars – moved in.

The planned Asda fleet fuel, like all commercial biodiesel, is completely legal but will still undercut conventional diesel prices by at least 10p a litre. Converting an in-house product like the waste oil will add to savings for the firm. 'Oil's a finite resource and we are fully aware of the fact that we shouldn't be wasting it,' Ms Fellows said. 'This is real eco-innovation – trials already show that chip pan fuel emissions are up to 40% lower that diesel.'

(Source: Wainwright, M., 29 Oct 2002, The Guardian)

Vehicle technology

Being liquids at room temperature, ethanol and methanol can be handled in a similar way to conventional fuels. Both have high *octane* ratings (enabling a high engine compression ratio which increases engine efficiency). They can be used in spark-ignition (petrol) engines with little or no modification as alcohol-petrol blends (e.g. E10 is 10% ethanol; also known as *gasohol*) or as pure alcohol fuels in modified vehicles. The suitability of alcohols as vehicle fuels is demonstrated by their use as high performance motor-racing fuels, for example in the US's 'Indianapolis 500'.

The principal difficulty with alcohol fuels is their relatively low energy density. This means that vehicles running on pure alcohol require a storage vessel double the volume of an equivalent petrol tank. Also, as alcohols are difficult to vaporise at low temperatures, pure alcohol vehicles are difficult to start in cold weather. For this reason, alcohol fuels are usually blended with a small amount of petrol to improve ignition. Methanol has the added disadvantage that it is both highly toxic and *hydrophilic* (mixes readily with water in all proportions) which can be a danger if used near to sources of potable water.

Alcohol fuels are already added to petrol to improve octane ratings and as *oxygenate* additives (to reduce carbon monoxide emissions). In the US alone, more than 1.5 billion gallons of ethanol are added to petrol each year (Johansson *et al.*, 1993). Although many other countries, including the UK, also use ethanol as a petrol additive, most consumers of petrol vehicles are unaware that this practice occurs. Methanol reacted with isobutylene to form methyl tertiary butyl ether (MTBE) is also used as an oxygenate additive. For example, over 60 kilotonnes are used as a fuel additive in the UK each year (which represents around 0.5% of petrol use by mass) (Hart *et al.*, 1999). However, the use of MTBE as a petrol additive is being discouraged due to new health concerns associated with its use.

Biodiesel is primarily used by heavy-duty vehicles as this sector is almost wholly dependent on diesel engine technology and very few alternatives exist for trucks of high tonnage. Most modern heavy-duty diesel engines can use biodiesel without modification provided the fuel is of the correct specification. Biodiesel can also be blended in any proportion with mineral diesel. One potential problem of 100% biodiesel fuel (B100) is an increase in the corrosion of rubber products. Engines and equipment with rubber seals and piping are usually replaced with non-rubber alternatives (a B5 blend does not lead to this problem).

As biodiesel has a lower energy density than mineral diesel, its use results in an increase in fuel consumption of around 5% (a B10 blend would result in a 0.5% difference). Existing fuel tanks therefore give slightly less mileage when using biodiesel. Another minor problem is that B100 is more viscous than mineral diesel in cold weather. However, a CFPP (cold filter plugging point) additive can alleviate this problem, enabling even pure biodiesel to be used in temperatures as low as –22 °C.

Fuel supply and infrastructure

Ethanol is one of the most widely used alternative vehicle fuels in the world due largely to its widespread use in Brazil and the US. Over 50 production plants in North America are in operation providing fuel ethanol production from starch crops (primarily corn). In the early 1990s, about 7 million tonnes of corn were used annually to provide more than 3 billion litres of ethanol for E10 alcohol blends, the gasohol equivalent of around 8% of the US petrol market.

During the 1970–80s, ethanol produced from sugar cane was vigorously promoted in Brazil both as a response to a slump in the global price of sugar and to reduce the country's dependence on foreign oil imports. At that time, all light-duty vehicles were required to run on at least in part on ethanol fuel. In 1989, the country's total fleet of cars and light-duty vans consisted of over 4 million pure-ethanol and 5 million gasohol vehicles. From 1973–87, even though the country's total energy demand almost doubled, petrol use dropped from 12% to only 4% of the energy market, while ethanol production increased to 18%. As a result of this and new home-production of oil and natural gas, the country's dependence on oil imports reduced by almost half. It is salutary to note, however, that the use of ethanol in Brazil sharply decreased due to ethanol shortages in 1989–90 which were caused by the price of ethanol being held artificially low (to hold down inflation) and the increasing availability of Brazilian oil and gas (Johansson *et al.*, 1993).

World annual demand for methanol is around 26 Mt of which 35% is used for formaldehyde production for use in the building industry in the manufacture of engineering board products. Of the remainder, 27% is for use in MTBE used as a vehicle fuel additive and 3% is used directly as a motor fuel. The UK demand is currently around 884 kt (24% formaldehyde, 7% MTBE), 550 kt of which is produced at the ICI Billingham plant (on Teesside) with the rest imported. Methanol is being considered as a possible vehicle fuel for future fuel-cell vehicles due to its suitable handling properties and ability to be *reformed* to hydrogen (see later section on *fuel-cell vehicles*). If this option is chosen for future vehicle design, global methanol demand could rise by orders of magnitude and would require new capacity to be built (Hart *et al.*, 1999).

Biodiesel is widely produced in Germany, France, Italy, and Austria. In 1994, following favourable signals from the EU, France set up four biodiesel plants which produced 100 kt/year, while a UK biodiesel consortium produced 18 000 litres of RME for pilot projects to assess its potential (Boyle, 1996). Assessments of land use have shown that if the maximum area of UK set-aside land was used for production of RME, this would generate around 700 kt/year (ETSU, 1996). This would provide an equivalent of around 5% of current diesel use. In the 2001 Budget, the UK government reduced the fuel duty on biodiesel by 20p/litre, indicating a high level support for the fuel in the near term.

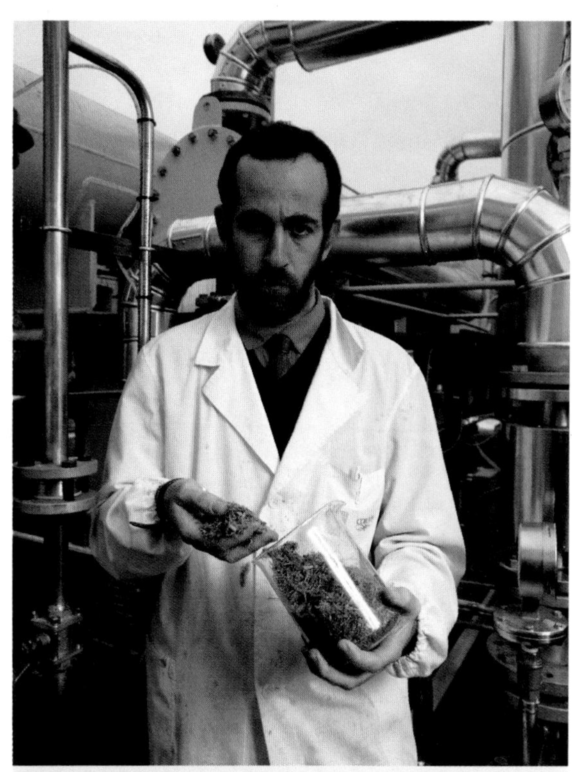

Figure 2.11 Biofuel crop being processed

Environmental impact

Although difficult to quantify, the consensus is that CO, HCs and particulates are reduced for M85 (85% methanol and 15% petrol), E85 blends and pure-alcohol fuels. Though alcohol fuelled vehicles can emit less nitrogen oxides (NO_x) (as alcohol fuels burn at a lower temperature than petrol), in practice the compression ratio is often increased to improve engine efficiency, increasing combustion temperatures and offsetting any reduction in NO_x emissions. Unburned alcohols present in the exhaust gases of an alcohol-fuelled engine contribute less to tropospheric-ozone formation than do the volatile organic compounds present in petrol exhaust emissions.

Comparing *regulated* emissions for biodiesel with current diesel types is made difficult due to the lack of back-to-back tests using ultra low sulphur diesel (ULSD) as a baseline (most data quoted uses *low sulphur diesel* which is 500 ppm sulphur). Estimates based on a number of comparative tests suggest that, for heavy-duty bio-diesel operation, particulate emissions are of the order of 10% lower than with ULSD. Bio-diesel's low sulphur content also allows the use of emission control systems (such as the CRT – see Section 2.5), which can further reduce particulates. However, without any emission control system, NO_x emissions can be increased for bio-diesel by 5–10%.

The great promise of biofuels is their *potential* to be carbon neutral, all the CO_2 emitted during processing and use of the fuel being balanced by the absorption from the atmosphere during the fuel crop's growth. However, in practice, unless organic growing methods are used, this is rarely the case, as the process of growing the biomass requires the input of fossil fuels for fertilisers, harvesting, crop processing and fuel distribution. The actual extent of total greenhouse gas emissions is therefore strongly dependent on the energy crop and the fuel processing used.

For example, in Brazil, where sugarcane is used as the feedstock for ethanol production, large amounts of *bagasse* (woody fibres remaining after the juice is extracted from the cane) are used to provide the process heat energy. As a result, the average energy-ratio of ethanol output to fossil fuel input is of the order of 6; i.e. six units of energy are produced for each unit input. Therefore, on a life-cycle basis, carbon emissions are significantly reduced. This contrasts with the net energy ratio for corn-derived ethanol from the US, which can in some cases be negative; i.e. the fossil fuel required to produce the ethanol is *greater* than the energy value of the final product.

The same wide variation in life-cycle CO_2 emissions is true for methanol as the emissions depend on the feedstock and processes employed. Although emissions associated with methanol from biomass can be lower than for conventional fuels, if fossil fuels are used as energy feedstocks, there is little difference between methanol and using petrol or diesel. Similarly, the results of life-cycle analysis of greenhouse gas emissions for biodiesel depend on the production processes employed. However, some studies show that, for RME, these emissions can be reduced by up to 50%, even when upstream emissions from the production of fertiliser are included in the analysis (Peterson and Hustrulid, 1998).

BOX 2.9 Pros and cons of vehicle biofuels

Advantages:

- **Reduced emissions** – can reduce life-cycle greenhouse gas emissions by up to 50%.
- **Reduced fuel costs** – current tax incentives provide low fuel costs per km for some biofuels.
- **Security of fuel supply** – useful alternative to importing of crude oil products.

Disadvantages:

- **Increased maintenance** – vehicles switching from conventional to biofuels may require extra servicing.
- **Land requirements** – large amount of land area required to supply existing vehicle fleet.

2.9 Battery electric vehicles

Battery electric vehicles (BEVs) are ideally suited to applications that benefit from zero-emission operation. These include use as small city cars, light-duty vans for freight delivery and industrial vehicles such as folk lift trucks that are used within an enclosed space. There are currently more than 28 000 BEVs in Europe, which includes over 15 000 milk delivery vehicles in the UK, one of the largest BEV fleets in the world. However, other than milk floats, there are only around 200 modern electric vehicles in use on British roads (EST, 2001a).

Vehicle technology

The design principle of a battery electric vehicle is relatively simple. Electrical energy (from any source of primary energy) can be stored in a 'secondary' or rechargeable cell on-board the vehicle. When required, electrical energy is drawn from the cells and converted to motive power by the use of an electric motor.

BEVs are significantly more energy efficient than conventional vehicles in stop-start traffic, as they use almost no energy when 'idling'. Electric vehicles can also recover the energy usually lost when braking via *regenerative* braking systems (up to 20% can be recovered). For these reasons, a battery with a *specific energy* (defined as the energy content per unit mass) of around 200 Wh/kg would provide a small BEV with a range comparable with a conventional passenger car.

The tried-and-tested lead-acid battery is the most widely used for BEVs. Although they have a relatively low specific energy (30–40 Wh/kg), it is possible to build a vehicle that has a range of around 70–90 km using lead-acid technology. Although these cells are far from ideal in their energy storage and power delivery characteristics, they are known for their reliability and durability, and are supported by an extensive maintenance network.

Other common traction batteries include nickel-cadmium (Ni-Cd), nickel metal-hydride (Ni-MH) and lithium-ion (Li-ION). Their higher energy density (50–90 Wh/kg) provides a significant improvement on lead-acid technology, increasing both vehicle performance and range. However, these battery types are expensive to produce and their use involves handling toxic materials such as cadmium and lithium. Despite these problems, these batteries have proved to be well suited to motive applications and are now preferred by many BEV manufacturers.

Most first generation BEVs used *direct current* (d.c.) motors that are cheap, give high torque at low speed and are easy to control using semi-conductor technology. However, their efficiency of 80–85% and *specific power* of 150–200 W kg^{-1} (about a third of that of a petrol engine) does not represent the best possible performance of available motor technology. One alternative is to use the a.c. induction motor, which has increased efficiency and double the specific power, but involves the use of a more costly control system.

The main disadvantage of BEVs is their high capital cost (typically an increase of 50–100%). For example, the Peugeot 106 Electric retails at £13 680 plus a battery leasing cost of around £70 per month (compared with around £8000 for the petrol version). In addition, most BEVs do not match the performance of conventional vehicles. However, current BEVs have a range and performance which is adequate for many specific, urban applications and are particularly suited to drive cycles that are predictable, regular and less than 160 km per day (e.g. delivery cycles) especially in areas where low emission vehicles are preferred or mandated. BEVs are therefore suited for use in commercial fleets (for small loads), company car-pools and within rental fleets.

Fuel supply and infrastructure

The most usual charging cycle is an overnight 'trickle-charge' from a standard domestic 13A, 230 V socket. This typically takes 6–8 hours and requires the use of a transformer to reduce the voltage and the current, which is then rectified to charge the cells using d.c. Fast charging (which takes less than 1 hour) is also possible using widely-available three-phase 400 V supply or a connection to an 11 kV high voltage line (found in many non-domestic and industrial buildings). Fast charging units require extensive electrical infrastructure (including transformer and switchgear) and are only cost effective when multiple points are being considered for a site.

Using the national electric grid, it is relatively easy and inexpensive to install slow-charge points as compared with other alternative fuels. Total installation costs per (standard) charge point are in the order of £500. However, fast charging systems (required for publicly accessible refuelling points) cost in the order of £7000–£30 000 per point (depending on type).

Most charging systems use a conductive cable to transfer the electrical energy to the vehicle. However, this is not the only option. *Inductive* (non-contact) charging systems are also being developed and have been successfully demonstrated in the French Praxitèle project. Yet another approach to vehicle recharging is to recharge the batteries away from the vehicle. In Birkenhead, UK, six Techobuses use battery packs that are recharged at the fleet's base. When refuelling is required, the depleted battery pack is exchanged for a fully charged pack.

BEVs have low fuel costs per km, due to the low price of electricity relative to other road fuels, and to the high efficiency of the electric drive-train. For example, a typical battery electric car costs less than 1p per kilometre to run (compared to a *fuel* cost of around 10 p/km for a petrol car) (Lane, 1998). Over an annual mileage of 15 000 km per year, this would represent a cost saving of around £1200 per year. However, once battery lease costs are taken into account (£70 per month), this saving is reduced to around £400 per year.

In spite of the lower combined battery and fuel costs, it is very difficult to accumulate the mileage necessary to recoup the extra capital required. For a Peugeot 106E, around 24 000 km would have to be covered per year averaging almost 100 km per working day (which exceeds the range possible on one charge). *Opportunity* charging during the day would therefore be required. Even with government subsidies, over the lifetime of the vehicle, the costs to the user are increased by 10–15% (DTI, 2000). Costs, therefore, remain a significant barrier to the introduction of BEVs.

Environmental impact

The battery powered electric vehicle is essentially a zero-emission vehicle *at the point of use*. Electricity used to recharge battery electric vehicles can be generated by the combustion of primary fossil fuels, the fission of nuclear fuels or can be produced using renewable sources. If renewables are used, a BEV can be operated with zero fuel-associated emissions on a *life-cycle* basis.

As electricity is currently produced from a range of energy sources (including coal, nuclear, oil, hydro and natural gas), we need to consider the production processes in detail if we are to be able to analyse the impacts of electricity use within the transport sector. Using the current fuel mix, the *life-cycle* data shows that CO, HCs and particulates are reduced for BEVs (as compared to petrol), although with an increase in sulphur emissions. NO_x emissions are significantly reduced for the heavy-duty BEVs, but can be higher when compared with light-duty petrol engined vehicles. For life-cycle CO_2 emissions of grid-electric fuelled BEVs, data shows a reduction in greenhouse gas emissions of up to 25% (compared to a diesel baseline). The trend is towards a generally cleaner electricity generating mix, with an increased fraction of Combined Cycle Gas Generation (CCGT) plant and renewables. Therefore, the local and greenhouse gas emissions associated with electricity generation (and BEV use) are reducing over time.

The benefits of BEVs to urban air quality are two-fold, lowering the overall emissions (gaseous and noise) and removing the emission sources from the street-level in city centres where the greatest number of people work and live. Furthermore, predicted life-cycle emissions reductions are often underestimated as the equivalent emissions for other fuels are based on hot engine conditions and do not account for cold start conditions when a high proportion of emissions from ICE vehicles can occur.

BOX 2.10 Pros and cons of battery electric vehicles

Advantages:

- **Zero-emission at point of use** – can also utilise renewable electricity so providing life-cycle zero-emission transport.
- **Reduced noise** – BEVs are almost silent at slow speeds and have low vibration in operation.
- **High efficiency** – electric drive system is more energy efficient than an ICE in stop-start driving.
- **Regenerative braking** – can recover up to 20% kinetic energy normally 'lost' in a conventional vehicle.

Disadvantages:

- **High capital vehicle cost** – high cost of electric drive-train and batteries can double the vehicle capital cost.
- **Limited vehicle range** – typical small BEV range of less than 100 km due to limitations of battery energy storage.
- **Long recharge time** – typically 6–8 hours for a slow charge.
- **Increased vehicle weight** – battery pack increases vehicle mass by 300–900 kg.

BOX 2.11 Nottinghamshire Millennium EV Project

Launched in Sept 1999, the Nottinghamshire Millennium Electric Vehicle (EV) Project brought 38 EVs (the largest number ever in a single UK project) onto the streets of Nottinghamshire. A total of 19 public and private organisations took part. The aim of the project was to promote green transport and reduce traffic-related air and noise pollution. Powergen, in association with PowerShift, formulated a project designed to attract high profile exposure and maximise its effectiveness.

The original funding came from Energy Efficiency Standards of Performance (EESoP) programme. EESoP was instigated by the electricity regulator, OFFER, as part of its 1994 review of supply regulation. The scheme was designed to provide quantifiable efficiency savings and to strengthen the UK's commitment to CO_2 reductions as agreed at the 1992 Rio Conference. Through this scheme all UK public electricity suppliers put aside £1 per customer, per year from April 1994 to April 1998, to be spent on energy efficiency activities.

Powergen and PowerShift initially drafted a programme of activities. Powergen approached public/private partnerships with an interest in developing energy conservation projects. Nottinghamshire County Council and Nottingham City Council's energy partnership provided a network of communication through which many varied organisations were contacted. In an effort to maximise funding potential, Powergen and PowerShift's initiative provided £300 000 for EV funding.

In September 1999, Powergen hosted an EV seminar designed to measure the general level of interest. Organisations interested in acquiring an EV within the Nottinghamshire area were invited to apply to PowerShift for funding before November 1999. Approved applicants received funding notifications by February/March 2000 with the majority of vehicles on the road before the end of May.

The objectives of the Nottinghamshire Millennium Electric Vehicle Project are to:

- reduce urban traffic pollution
- promote the use of clean fuel vehicles and raise awareness of environmental issues
- show that electric vehicles really are a practical option for everyday use
- encourage the use of renewable energy resources for transport consumption
- involve the wider community and raise awareness regarding clean fuel vehicles

Peugeot and Citroën supplied the vehicles used in the project and therefore were responsible for ensuring adequate technical support locally. There were a few initial problems connected with delays in vehicle delivery and colour specifications but, generally, those taking part in the project have been pleasantly surprised with the reliability and simplicity of their EVs.

The Peugeot 106 EV has a top speed of 56 mph and [a] 50-mile range. The vehicle's electronic control allows energy to be recovered through regenerative braking. Batteries can be recharged anywhere using a standard 240 V 13 amp socket connected to the vehicle's on-board charger. Maximum recharge usually takes six hours although a one-hour recharge will provide approximately 13 miles of travel. Using Economy 7 off-peak tariff, a full battery recharge costs less than 50p, or approximately 1p a mile over an average 50-mile radius.

EVs are exceedingly quiet and local air emissions are negligible. Vehicle emission savings compared to petrol/diesel alternatives are considerable. Estimates show that even when fossil fuels are used to generate electricity, 18% less carbon dioxide (CO_2) is produced through the use of EVs. Alan Allsopp, principal energy management officer at Nottinghamshire County Council stresses that: 'We are intending to recharge the vehicles from green electricity produced from renewable sources.' Electricity is the most efficient fuel for powering vehicles in congested traffic. Internal combustion engines continue to burn fuel whilst stationary – EVs only use energy when moving. As a result, EVs consume approximately 30% less energy than conventional vehicles.

In most cases PowerShift funded the basic price difference between a conventional vehicle and EV, while Powergen provided matching funding to cover the battery lease cost for the first five years. This enabled organisations to purchase EVs at the same price as an equivalent petrol/diesel vehicle. Operating savings are estimated to be £5000 per vehicle, on average, over a five-year period. Nottingham Energy Partnership representative Ravi Subramanian believes: 'All too often energy partnership groups struggle to deliver tangible benefits to the wider community. That was certainly not the case on this occasion.'

The project provided the perfect opportunity for organisations to put environmentally sustainable policies to the test. The EESoP scheme is primarily aimed at reducing CO_2 levels, but this project also has substantial air quality benefits for Nottinghamshire. Steve Waller, sustainability team leader for Nottingham City Council commented: 'This was a great initiative – the project was very well promoted and targeted the right people for the right reasons.' Paul Edwards, energy efficiency project manager at Powergen, said: 'We were delighted with the take-up of the project and the very positive feedback we have received from participants now they are using the electric vehicles. As the largest electric vehicle project in the UK, it is delivering considerable environmental benefits and pointing the way for others.'

Source: EST, 2001b

2.10 Fuel cell electric vehicles

> There seems to be a feeling creeping through the motor industry that perhaps the days of the internal combustion engine are numbered
>
> (Hart and Bauen, 1998a, p. 7).

As discussed in Book1, Chapter 14, there are many reasons to support the transition from a carbon-based energy system to a *hydrogen economy*. Primarily, this is to reduce overall carbon emissions, which are associated with climate change. However, the use of hydrogen as a fuel also provides other advantages. For example, hydrogen gas has the highest energy-to-weight ratio of all fuels, with 1 kg of hydrogen containing the same amount of energy as 2.5 kg of natural gas or 2.7 kg of petrol (see Tables 2.1 and 2.5).

The use of hydrogen as a fuel is not new. 'Coal gas' or 'town gas', which is at least 50% hydrogen, has been used extensively throughout the industrial nations and preceded the use of natural gas in North America and Europe. As noted in Book 1, Section 14.5, around 1.5% of world energy supplies are already converted to hydrogen gas for use in the chemical and petrochemical industries. The gas is currently used for the chemical synthesis of ammonia, ethylene and methanol and in the desulphurisation and hydrogenation of fossil fuels.

Hydrogen is a versatile fuel and can be used within modified internal combustion engines (see Book 1, Section 14.5.1). Since the 1970s, BMW have developed a series of hydrogen ICE prototype cars, the latest of which include the bi-fuel 750hL and the hydrogen-powered Mini. The only combustion products from a hydrogen powered ICE are water vapour and small amounts of NO_x (due to the presence of atmospheric nitrogen). In addition to reduced emissions if used within an ICE, the use of hydrogen as a fuel also offers the possibility of using an alternative engine technology, the *fuel cell*.

The fuel cell

Fuel cells are electrochemical devices that convert chemical energy directly into electrical energy, heat, and water. The principles of fuel cells are similar to those of electric batteries, where energy conversion is taking place between the reactants to produce electricity. However, unlike a battery, a fuel cell does not store chemical energy. The reactants (fuel and oxidant) have to be continually supplied to the cell for an electric current to be produced.

The anode and cathode of a fuel cell are separated by an *electrolyte* that allows the transfer of ions, but physically separates the fuel and oxidant. This prevents the exchange of electrons that would be required for a non-catalytic chemical reaction to occur (i.e. combustion). When the reactants are fed into the cell, chemical reactions take place. The main charge carriers (usually H^+) cross the electrolyte and the electrons are transferred via an external circuit. The electric current produced can be used to drive a motor or other external load. The fuel normally used is hydrogen or a hydrogen-rich compound (supplied to the anode) and the oxidant can either be pure oxygen or air (supplied to the cathode).

Single cells typically generate around 0.8 V with a power output of up to 100 W. Larger outputs are achieved by assembling cells in series or parallel to form a *stack*, which has the required voltage and output characteristics. In contrast to heat engines, fuel cells are not limited by the Second Law of Thermodynamics, which means that they are able to achieve higher conversion efficiencies than heat engines (this was first covered in Book 1, Chapter 6). Although there are losses within a fuel cell that arise due to ohmic resistance of the cell components, efficiencies of up to 80% have been demonstrated in the laboratory.

BOX 2.12 Principle of operation of a PEM fuel cell

The PEM cell uses highly conducting electrodes made of graphite, which form the terminal of each cell and separate adjacent cells in the stack. The electrodes are grooved to allow easy passage of the gases to the 'surface of action' while also maintaining electrical contact with the electrolyte-catalyst-gas interface. At the anode, hydrogen is catalytically disassociated to leave hydrogen ions. An external circuit conducts electrons while the positive ions (protons) migrate through the electrolytic membrane to the cathode. There they combine, again under action of a catalyst, with oxygen and electrons returning from the external circuit, to form water

Several fuel cell types have been developed, each being characterised by the electrolyte used, operating temperature and fuel gas quality required (refer back to Book 1, Table 14.3 for types of fuel cell compared). Low-temperature fuel cells include the Alkaline Fuel Cell (AFC), the Solid Polymer Fuel Cell (SPFC), of which there are two types: the Proton Exchange Membrane Fuel Cell (PEMFC), and the Direct Methanol Fuel Cell (DMFC). Due to the relatively low temperatures within the cells, these usually require a catalyst at the anode to promote the necessary reactions taking place. High temperature fuel cells (650–1000 °C) include the Molten Carbonate Fuel Cell (MCFC) and the Solid Oxide Fuel Cell (SOFC).

Vehicle technology

If fuel cells are to replace the internal combustion engine in road vehicles, they need to have comparable power and a similar response time. In practice, this means a power density of at least 1 kW kg^{-1} (for cars) and a start-up time measured in seconds. The fuel cell thought by most analysts to meet these requirements is the PEM fuel cell, which has the ability to operate at relatively low temperatures, so reducing start-up times. Solid polymer electrolyte materials such as Nafion (related to Teflon) also eliminate the safety considerations associated with liquid acid and alkali electrolyte cells.

Following the successful use of fuel cells in the Gemini and Apollo space missions (which used solid polymer and alkaline fuel cells respectively), the 1960s saw a number of terrestrial fuel-cell vehicle prototypes. These included Shell's 20 kW fuel cell truck and General Motor's liquefied hydrogen-oxygen fuelled Electrovan, which was powered by a 5 kW Union Carbide fuel cell. In the 1970s, interest in fuel cells was renewed due to the sharp increase in world oil prices and the decade saw designs such as the hybrid AFC-battery Austin A40 car, which used roof-mounted compressed hydrogen tanks and had a range of 300 km (Hart and Bauen, 1998a). However, interest in road fuel-cell vehicles declined in the 1980s as the fuel crises of the 1970s receded.

In the 1990s interest in fuel cells for road transport was revived, this time with a focus on the environmental benefits that the technology could provide. This decade saw the development of fuel-cell vehicle (FCV) prototypes by most of the major vehicle manufacturers and the emergence of new companies specialising in the manufacture of fuel cell systems. One such company is Ballard Power Systems who, in collaboration with DaimlerChrysler and Ford, developed the world's first fuel-cell bus and the Necar (New Electric Car). One of the aims of the Ballard-DaimlerChrysler-Ford partnership is to market a fuel cell car, for use by the general public, by 2004.

Figure 2.12 The Necar fuel-cell vehicle

A Necar prototype was built for the California Fuel Partnership based on the Necar 4. The objective was to test the car in real driving conditions over 40 000 km on the streets of California.

The Necar programme was specifically designed to develop a commercial PEM fuel-cell vehicle. From 1994 to 2000, five prototypes were tested (Necars 1 to 5) with the objective of reducing the mass and volume of the fuel cell stack and on-board fuel system to a size suitable for passenger car applications. During the programme, the stack power output improved from 5 kW to 75 kW. Three on-board fuel systems were also tried; Necars 1 and 2 used compressed hydrogen, Necar 4 was fuelled with liquid hydrogen and Necars 3 and 5 used methanol and an on-board *reformer* to generate hydrogen on demand (see next section). The Necar 5 achieved a top speed of 150 km h^{-1} and a range of 400 km on an 11-gallon tank of methanol (HyWeb, 2001).

DaimlerChrysler also developed the Nebus, a fuel cell version of a 70-passenger single-deck bus. High pressure (30 MPa) hydrogen cylinders mounted on the roof were used to power ten 25 kW PEM fuel cell stacks providing a range of 250 km. The Nebus was initially tested in Germany and North America (in Chicago and Vancouver). These successful demonstrations were followed by a development of the ZEbus (Zero Emission Bus), which was used as part of the California fuel cell Partnership programme. Further trials include the Clean Urban Transport for Europe (CUTE) fuel cell bus programme, which is demonstrating 30 fuel cell 'Citaro' buses in several European countries during 2002–2003.

Hydrogen's low density has presented a technological challenge to the design of on-board hydrogen storage systems. At room temperature and pressure, to store an equivalent amount of energy to that contained in a typical petrol tank would require a hydrogen tank with around 800 times the volume. From a technical perspective, three main methods of on-board hydrogen storage are under consideration (as of 2003). These are *compressed* gas, *liquefied* gas and *metal-hydride* storage.

Compression is the least expensive of the three options, the gas being stored in cylinders at pressures up to 30 MPa. The most advanced tanks, which incorporate lightweight materials such as aluminium and carbon fibre, can achieve up to 3.6 MJ kg^{-1} (this compares with 32 MJ kg^{-1} for petrol plus tank). Cryogenic systems, which store hydrogen in its liquid state at low temperature (–253 °C), can achieve an energy density of around 16 MJ kg^{-1}. However, ultra low temperature systems are expensive and liquefaction requires large amounts of energy, the energy required being about 40% of the energy stored. The third storage option is to use metal-hydrides that absorb hydrogen when under pressure, the gas becoming part of the metal's physical structure. The advantages of hydrides are the low loading pressures (less than 10 MPa), ease of use and high level of safety. However, hydrides are limited by their low energy density (up to 1.4 MJ kg^{-1}) and the complexities of the refuelling equipment (HyWeb, 2001).

Other methods of hydrogen storage being developed include the use of *carbon adsorption* whereby hydrogen is adsorbed by carbon nano- or micro-fibres under pressure. Initial evidence suggests that this technique could enable storage densities higher than are achieved with liquefaction. However, these technologies remain at the development stage (Bérnard and Chahine, 2001).

Fuel supply and infrastructure

While there is a high level of agreement regarding which type of fuel cell is most suitable for road transport applications, the same cannot be said for the fuel supply system. This is primarily because of the large number of energy conversion routes that can be used to deliver hydrogen to the fuel cell. (Note that, like electricity, hydrogen is a secondary fuel and must therefore be produced from primary energy sources.) However, the large number of production routes is also one of the great strengths of the hydrogen economy as the gas can be produced from almost any primary energy source.

Large-scale processes developed for the production of hydrogen from fossil fuels are well established. Currently, over 80% of hydrogen production is sourced from natural gas using the process of *steam reforming*. Other carbonaceous feedstocks (such as, methanol, ethanol, and biomass) can also be used to generate hydrogen via processes that include *thermal decomposition, partial oxidation*, and *gasification*. Even water can be *electrolysed* to produce hydrogen, though it should be noted that energy is required to 'split' H_2O into its constituent elements, and therefore water should not be considered as the 'fuel'. If renewable energy in the form of electricity or biomass is used to produce hydrogen (via *electrolysis* or *gasification*) and used in a FCV, this could potentially provide road transport with zero-emissions (apart from the production of water vapour) on a lifecycle basis.

The main energy conversions leading to hydrogen can be categorised according to the location of hydrogen production, which can occur in one of three ways. Firstly, hydrogen can be produced centrally and then distributed to fuel stations where it is compressed and stored ready for use by an FCV. As is the case with large-scale production methods, a great deal of experience has been accumulated regarding hydrogen distribution on an industrial scale. For example, hydrogen is routinely transported by road in compressed form using steel bottles at 20 MPa, liquefied hydrogen is carried using 5000 litre capacity road-tankers and hydrogen gas is also routinely piped under pressure. There are already around a dozen hydrogen vehicle refuelling stations worldwide which including public access stations in Hamburg and at Munich airport (as shown in the video which accompanies Unit 23) and depot-based facilities supporting the fuel cell bus fleets in Chicago, Vancouver and California (HyWeb, 2001).

In the second category, a hydrogen carrier fuel is produced at a central location and distributed to fuel stations where it is processed to produce hydrogen on-site. (A hydrogen carrier fuel is defined here as any fuel which can be used to generate hydrogen on demand.) In the UK, natural gas could be used to generate hydrogen on-demand using small-scale reformers located at fuel stations. This would use the extensive natural gas grid that already covers a large proportion of the country. Similarly electricity from the national grid could be used to electrolyse water to produce hydrogen where and when required. Many analysts have proposed this option as the most cost-effective methods of hydrogen fuel infrastructure development as it makes use of the existing fuel infrastructure to maximum effect (Hart *et al.*, 2000).

Thirdly, a hydrogen-rich fuel can be processed on-board the vehicle (via catalytic reforming) so generating hydrogen gas on-demand. This approach avoids the problems of hydrogen storage already discussed and reduces the need to build new fuel infrastructure, which would be required for a nationwide hydrogen gas network. In principle, any hydrogen carrier can be used for this option, although simpler hydrocarbons are easier to reform. As on-board reformers need to have fast response times, fuels that can be processed at low temperatures are preferred. Of the liquid fuels, methanol is unique in that it can be reformed into hydrogen at around 260 °C, as compared to 600–900 °C for other fuels such as petrol, ethanol, natural gas, and propane. Therefore methanol is considered to be the prime candidate for on-board fuel storage and has been successfully demonstrated in test vehicles including the Necar 3 and Necar 5 FCVs. However, many manufacturers are attempting to develop petrol reformers, driven by the possibility that FCVs may be able to utilise existing fuel infrastructure.

Environmental impact

Precise vehicle performance data is difficult to source due to the current commercial sensitivity of fuel-cell vehicle development. However, what data is available suggests that fuel economy is significantly improved for FCVs as compared to conventional vehicles due to the high efficiency of the fuel cell drive-train in comparison with the ICE. Figure 2.13 shows estimates of energy use for a small FCV as compared to a petrol car. These figures are based on published test data for the Necar 4 and Ford's P2000 hydrogen car and also on computer modelling conducted by Hart. Note that the large range of estimates is a reflection of a higher degree of uncertainty for the less tested technologies. Hart also predicts a similar reduction in energy use for a hydrogen fuel cell bus as compared to its diesel equivalent (Hart *et al.*1999).

If vehicle energy use and fuel production emissions data are combined, *life-cycle* greenhouse gas emissions are predicted to reduce, due to improved efficiency of the vehicle and fuel processing. However, the reduction is difficult to quantify, depending as it does on the method of fuel production. If natural gas is reformed on-site at fuel stations, then modelling by Hart and others suggests that greenhouse gases will be reduced by almost 60% on a lifecycle basis for light-duty vehicles. Similar reductions are expected for a fuel cell bus. In principle, the use of renewable hydrogen would eliminate the emission of greenhouse gases altogether. However, this is likely to be an expensive option in the near-term.

Estimates for *regulated* emissions suggest very low emissions associated with fuel-cell vehicles use. Using a life-cycle analysis, and compared to an ICE baseline, Hart estimates that all four regulated emissions are likely to be reduced by at least 70% for methanol and petrol fuelled FCVs and by at least 85% for FCVs using compressed hydrogen produced from on-site reforming of natural gas.

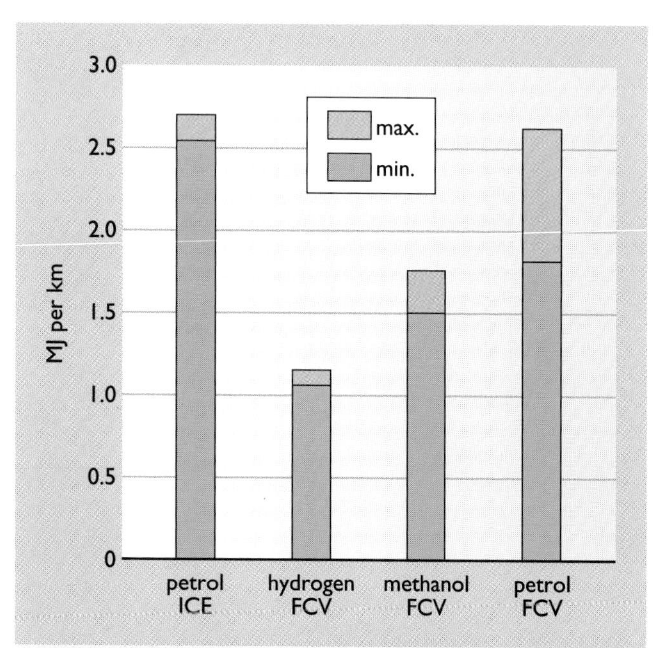

Figure 2.13 Comparison of vehicle energy use for pre-production fuel cell cars. (sources: data based on published test data for the Necar 4 and Ford's P2000 hydrogen car and also on computer modelling conducted by Hart (AEA, 1999, Hart, 1999, DTI, 2000, Thomas et al., 2000))

As noted in Book 1, Section 14.5.2, safety concerns could act as a barrier to hydrogen fuel-cell vehicle introduction. However, there is a growing body of evidence to support the view that the use of hydrogen is no more dangerous than the use of petrol or other flammable fuels. In hydrogen's favour, the gas is non-toxic. A hydrogen fire produces no poisonous fumes and has a lower flame temperature than petrol fuelled fires. Due to hydrogen's low density, escaping gas rises away from a spill site, unlike petrol vapour (and LPG) which remain in the spill area so prolonging the fire's duration. With a high diffusion coefficient, hydrogen mixes in air faster than petrol vapour or natural gas, which is advantageous in the open (but could represent a potential disadvantage in a poorly ventilated enclosed space). The researcher Kuhn states that in extensive tests of pressurised hydrogen cylinders: 'Such tanks were crashed, crushed, dropped, shot, burned, and blown up, but failed to produce any consequences as bad as those resulting from comparable assaults on ordinary gasoline tanks' (Williams, 1997).

Taken overall, while the risks of using hydrogen may not be greater than using conventional fuels, its use as a vehicle fuel requires *different* handling procedures. For example, hydrogen is colourless and odourless which makes human detection difficult. Also, the gas burns in air at concentrations of 4%–75% by volume (which is a larger range than for other fuels) and the minimum ignition energy required for a *stoichiometric* hydrogen/oxygen mixture is only 20 μJ, one order of magnitude less than for natural gas and petrol vapour (Hart, 2000). The use of hydrogen also causes embrittlement in some metals, which must be replaced (usually by stainless steel) to prevent premature pipe breakage. Also, if methanol is used as an on-board fuel, handling precautions need to be taken, as methanol is highly toxic, can corrode some materials and is *hydrophilic* (which raises concerns of methanol spills near to supplies of drinking water). However, solutions to all these potential problems have been successfully demonstrated in many FCV demonstrations worldwide.

References

AEA (1999) *Automotive Environment Analyst*, London, Financial Times.

AFDC (2003) Alternative Fuels Data Center website, URL: http// www.afdc.doe.gov/ pdfs/fueltable.pdf [accessed 14 May 2003].

Automotive News Europe (2001) Volume 6. No. 1, Page 16. 1 Jan. 2001, Crain Communications Inc., London, UK.

Bérnard, P. and Chahine, R. (2001) Modelling of adsorption storage of hydrogen on activated carbons, *International Journal of Hydrogen Energy*. vol. 26, pp. 849–55, Elsevier Science.

Boyle, G. (Ed.) (1996) *Renewable Energy, Power for a Sustainable Future,* Milton Keynes, Open University in association with Oxford, OUP.

Cuddy, M. and Wipke, K. (1997) *Analysis of the Fuel Economy Benefit of Drivetrain Hybridization*, USA, National Renewable Energy Laboratory.

Davis, S. C. (2001) *Transportation Energy Data Book*, Center of Transportation Analysis, Edition 21, Tennessee, Oak Ridge National Laboratory.

DTI (2000) *The Report of the Alternative Fuels Group of the Cleaner Vehicle Task Force Report*, Department of Trade and Industry, Automotive Directorate, London, The Stationery Office. Available from FTP: http:// www.roads.detr.gov.uk/cvtf/index.htm [accessed 14 May 2003].

EST (2001a) *PowerShift Market Survey*, London, Energy Saving Trust.

EST (2001b) PowerShift website. URL: http://www.est-powershift.org.uk/ [accessed 10 June 2003].

ETSU (1996) *Alternative Road Transport Fuels: A Preliminary Life-cycle Study for the UK, Volume 2*, London, ETSU, The Stationary Office.

Hart, D. and Bauen, A. (1998a) *Fuel cells: clean power, clean transport, clean future*, Financial Times Report, Financial Times Energy.

Hart. D. and Bauen. A. (1998b) *Further assessment of the environmental characteristics of fuel cells and competing technologies*, ETSU, UK (ETSU F/02/00153/REP).

Hart, D., Fouguet, R., Bauen, A., Leach, M., Pearson, P. Anderson, D. and Hutchinston, D. (1999) *Methanol supply and its role in the commercialisation of fuel cell vehicles*, ETSU, UK (ETSU F/02/00142/REP).

Hart, D., Fouguet, R., Bauen, A., Leach, M., Pearson, P. and Anderson, D. (2000) *Hydrogen supply for SPFC vehicles*, ETSU, UK (ETSU F/02/00176/ REP).

HyWeb (2001a) HyWeb website, URL: http://www.HyWeb.de/ [accessed 10 June 2003].

IEA (1999) *Implementation Barriers of Alternative Transport Fuels*, IEA/ AFIS, Innas BV, February 1999.

IEA (2002) http://www.iea.org/statist/keyworld2002/key2002/keystats.htm [accessed 10 June 2003].

Pearson J. K. and Kapus P. (2000) *Gasoline engine concepts related to specific vehicle classes*, IMechE Conference Transactions, Int'l Conf., 21st Century Emissions Technology, paper no. C588/009/2000, ISBN 1 86058 322 9, Bury St Edmunds and London, Professional Engineering Publishing Limited,. pp. 211–227.

Kitman, J. L. (2000) 'The Secret History Of Lead – A Special Report', *The Nation*, March 20, 2000.

Johansson, T. B. *et al.* (ed.) (1993) *Renewable Energy: Sources for Fuels and Electricity*, London, Earthscan, Island Press

Lane, B. (1998) *Promoting electric vehicles in the United Kingdom: A study of the Coventry Electric Vehicle Project*, case study for the project 'Strategic Niche Management as a tool for Transition to a Sustainable Transport System', Milton Keynes, The Open University.

Monaghan (1998): 'Particulates and the diesel – the scale of the problem' (paper S491/001/98), *Diesel Engines – Particulate Control*, IMechE Seminar Publication, p. 16.

Mildenberger, U. and Khare, A. (2000) 'Planning for an environment-friendly car', *Technovation* 20, pp. 205–14.

Peterson, C. L. and Hustrulid, T. (1998) 'Carbon cycle for rapeseed oil biodiesel fuels', *Biomass and Bioenergy*, vol. 14, no. 2, pp. 91–101.

Serfas, J. A., Nahmias, D. and Appleby A. J. (1991) 'A Practical Hydrogen Development Strategy', *International Journal of Hydrogen Energy*, 16, pp. 551–6.

Standard and Poor (1998) *World Car Industry Forecast Report*, Lexington USA, McGraw Hill Companies, pp. 89–94.

Teufel, D., Bauer, P., Lippolt, R. and Schmitt, K. (1993) *OeKO-Bilanzen von Fahrzeugen*, 2. erweiterte Auflage, Heidelberg: Umwelt-u. Prognose-Institut Heidelberg e.V.

The Open University (2000) T172 *Working with our Environment: Technology for a Sustainable Future*, Theme 2: *Travelling Light*, Milton Keynes, The Open University.

Thomas, C. E., James, B. D., Lomax F. D. and Kuhn I. F. Jr. (2000) 'Fuel options for the fuel cell vehicle: hydrogen, methanol or gasoline?' *International Journal of Hydrogen Energy*, vol. 25, pp. 551–67, Elsevier Science.

Williams B. D. (1997) 'Hypercars: Speeding the transition to solar hydrogen', *Renewable Energy* vol. 10, no. 2/3, pp. 471–9, UK, Pergamon. Adapted from Lovins *et al.* (1996) Hypercars: Materials, Manufacturing and Policy Implications, Colorado, Rocky Mountain Institute.

Chapter 3

Mobility Management

by Stephen Potter and Marcus Enoch

3.1 Transport impacts and institutions

In Chapter 1, it was concluded that the only potentially viable approach to dealing with the energy and environmental impacts of transport is to combine technical improvements in fuel efficiency and lower carbon fuels with changes in people's travel behaviour. This follows the concept of 'intelligent consumption' – the idea that the benefits achieved by travel can be delivered at a lower energy and resource cost. Chapter 2 looked at technical methods of reducing emissions of pollutants, improving fuel efficiency and introducing lower carbon fuels. In this chapter and Chapter 4 we will look at the changes that can be made in our patterns of travel to reduce energy use and environmental impacts. This is not to say that technology has no role in adapting travel behaviour. Indeed, the development of certain key technologies is crucial to such approaches working. Adapting travel behaviour frequently requires technologies to support it, just as improving fuel efficiency and moving to lower carbon fuels may require some behavioural change on the part of vehicle users.

BOX 3.1 Transport terminology

The concept of managing the demand for transport is known by several names. In the USA the term *transportation demand management* (frequently abbreviated to TDM) is used. In Australia and the UK the variant *transport* or *travel demand management* is used. Among EU policy makers, *mobility management* is the most common term, and is coming into use in the UK as well. This is the term we shall adopt here. All these terms are largely interchangeable, although they may contain differences in emphasis.

Although transport policy is often viewed as something done by Government that affects individuals, it is becoming increasingly important for institutions such as employers, shopping centre managers and big service providers to have a role in transport policy. This is not an easy policy area, as many people are reluctant to accept what they see as infringements upon travelling in the way they desire, and organizations do not see it as part of their business to 'interfere' with the travel behaviour of their staff. Despite this, the role of employers and other institutions in supporting more sustainable transport policies is a new and important field. An example of an institutional response was given in Chapter 2 (Box 2.7), which described the introduction by Oxfordshire Mental Healthcare NHS Trust of cleaner LPG vehicles. However, rather than looking at cleaner vehicle technologies, this chapter and the next will focus on the potential for institutions to adjust their travel behaviour.

Figure 3.1 Employers locating in profitable city-edge locations have major transport impacts. Increasingly they are being asked to take some responsibility for the traffic and environmental effects generated

3.2 The 'bed of nails' of mobility management

The transport policy challenge

Back in 1981 the development of an 'integrated transport policy' was the subject of an episode of the BBC political comedy *Yes Minister*, entitled 'The bed of nails'. The plot was that transport involved so many irreconcilable desires, interests and approaches that the only politically viable transport policy was not to have a policy at all, least of all an integrated transport policy. More than twenty years later that has proved to be an astute observation. At the heart of the transport crisis is a widespread paradoxical reaction. While, on the one hand, people accept that our high and growing level of road transport dependence is costly, unpleasant, unsustainable and generally undesirable, they resist policies that actually affect themselves, their town or their company.

Behind this is a tension between collective effects and individual benefit. While increased reliance on cars has caused all sorts of problems for society as a whole, on an individual level the use of the car confers substantial benefits. Convenience, comfort, flexibility, personal space, and low perceived cost are often cited as reasons for the dominance of travelling by car, but other deeper reasons have been suggested.

> The sensual, erotic, or irrational well springs of the auto mobility cannot be ignored. The pleasure, as well as the convenience that auto driving provides is a boon to many people. However, what is needed is a transport system that allows people to find pleasure in many ways of travel. New policies must be as non-punitive as possible in discouraging auto use, and must develop seductive, as well as affordable and efficient alternatives to the auto.
>
> Freund and Martin, 1993

Figure 3.2 Car advertisements often play on the 'sensual, erotic, or irrational well springs of auto mobility'

Furthermore, for many people there is no real viable alternative to using their car. We have gradually adjusted our lifestyles over the years so that we depend on the sort of personal mobility that a car provides. The car has generated journeys and a lifestyle for which public transport is not nearly as convenient (if even possible), while walking and cycling are not feasible because distances for many journeys are now too great.

The crux of the problem is that the benefits of car use are very evident to individuals, whereas the problems are more diffuse and affect the population as a whole, with some impinging on future rather than current generations. This means that any individual action to reduce car use produces little or no obvious personal benefit. For example, one parent letting the children walk to school will not improve their safety, so long as no similar action is taken by other parents. Furthermore, the cut in pollution will seem negligible and any benefits may not be felt for many years. Any global benefits may not even affect the original country. This unequal conflict between choosing immediate and tangible personal benefit over a delayed, dispersed and far less visible cost to society is behind many of the difficulties faced in addressing the transport crisis. It is thus not surprising that transport policy is, politically, a bed of nails!

Targeting the 'easy wins'

Mobility management policies, although necessary for a variety of reasons, are thus very far from being politically popular. Policy makers therefore feel they need to go for the 'easy wins' and target those people most likely and able to alter their car use and to produce the most obvious benefits. An example may be park-and-ride sites at the edge of historic cities such as York, Oxford and Durham. There is an acceptance that traffic adversely affects the quality of life in the historic centre of these cities and that the narrow city-centre roads cannot be widened. The park-and-ride schemes target shoppers and commuters, for whom the change in behaviour is relatively easy.

The identification of appropriate groups of people or types of journey can be done in a number of ways. One method is to use socio-economic data to identify the type of person most likely to walk, cycle or use public transport, in a similar way to supermarkets' use of market research companies to match people to products. This technique is now slowly being adopted by the bus industry in Britain and in various towns and cities across the world (including Perth, Western Australia and Leeds, West Yorkshire). Termed *travel blending*, it involves identifying and 'educating' those most capable of switching from the car to other modes of transport. Box 3.2 (overleaf) provides more information about this approach.

Another approach is to identify not the people, but the reasons for the journeys being made. This involves considering why people make particular journeys and then trying to devise ways of reducing certain types of trip. For example, it may be possible to reduce the number of shopping trips by car through introducing teleshopping, internet shopping or home delivery schemes. The number of school journeys made by car might be cut by opening safe cycle routes, operating school buses or organizing supervised walking groups (called 'walking buses').

BOX 3.2 Travel blending

Travel blending – What is it?

Travel blending® is the terminology used to describe a way for individuals to reduce the use of the car which involves:

- thinking about activities and travel in advance (i.e. in what order can I do things, where should they be done, who should do them?) and then:
- blending modes (i.e. sometimes car, sometimes walk, sometimes public transport etc);
- blending activities (i.e. doing as many things as possible in the same place or on the same journey); and
- blending over time (i.e. making small sustainable changes over time – once a week or once a fortnight)

People and households who take part in travel blending® choose to change their behaviour by:

- observing their own current travel patterns – measuring the way they and their households use the car for one week;
- receiving detailed suggestions customised to those travel patterns;
- setting their own targets;
- spending some weeks trying to reduce the use of the car;
- observing the changes they have achieved; and
- being given a simple, ongoing system of monitoring and motivation.

Travel blending – Where has it happened?

Adelaide
Nottingham
Sydney
Leeds
Santiago
London
Hastings
Bristol
New Jersey

Reductions in car use for those participating have ranged from a 6% to a 23% reduction in car driver trips. A pilot in Nottingham gave the following results for participants ... who completed [travel] diaries and for the whole project population including non-participants ([including] those who were approached and refused to take part or dropped out at any time).

	Diary 1	Diary 2	% Change of participants	Change inc. % whole population
Car driver trips	19.1 trips	17.7 trips	–7.6%	–3.3%
Car driver miles	147.3 miles	126.5 miles	–14.2%	–6.2%
Total hours in the car (all resp.)	7.5 hours	6.6 hours	–11.8%	–4.8%

Steer Davies Gleave, no date

In theory, some of the easiest journeys to deal with are those that people make every day, e.g. commuting trips to and from work. These usually have a fixed start and end point (home and work), and are generally made at similar times each day. Commuting and business trips account for around one fifth of the total number of journeys made in the UK, currently amounting to some 11.4 billion trips each year. Of these, around 70% are made by car (Department for Transport, Local Government and the Regions, 2001). Any change in such behaviour therefore would make a sizeable impact on the transport problem.

Targeting commuter trips will frequently require the involvement of employers. So what role can employers play in changing the travel behaviour of their staff to reduce environmental impacts? In the USA from the late 1970s transport demand management (TDM) by employers became part of a range of initiatives to improve air quality. It was implemented by regulations requiring employers to cut car commuting (particularly single-occupant car commuting) to their sites. This approach is now developing in Europe, with a mixture of regulations, tax incentives and voluntary agreements seeking to stimulate employers to introduce measures to help their staff commute in a more environmentally friendly way.

BOX 3.3 More on transport terminology

It has already been noted in Box 3.1 that the concept of managing the demand for transport is known by several names, and in this text the term *mobility management* is mainly used. Within the general policy approach of mobility management, there are a number of measures, including employer-led initiatives. Such initiatives have also been referred to by a variety of terms, including *commuter plan, green commuter plan, mobility plan* and *green transport plan*. In the UK, the term *travel plan* is now the most common term, and we shall use it in this chapter and in Chapter 4, although you should be aware of the varying terminology. Mobility management and travel plans are new areas and the terminology has yet to settle down.

3.3 What is a travel plan?

The concept of a travel plan is central to the role of institutions in reducing traffic congestion, accidents and the environmental impacts of the transport activities of their staff, customers and visitors. Government guidance defines a travel plan as being:

> a general term for a package of measures tailored to meet the needs of individual sites and aimed at promoting greener, cleaner travel choices and reducing reliance on the car. It involves the development of a set of mechanisms, initiatives and targets that together can enable an organisation to reduce the impact of travel and transport on the environment, whilst also bringing a number of other benefits to the organisation as an employer and to staff.
>
> Energy Efficiency Best Practice Programme, 2001, Section 1.1

The crucial point about travel plans is that those organizations responsible for creating the need to travel, such as employers, service providers and shopping centre owners, are involved in helping to solve transport problems. The involvement of such institutional players is both a strength and a weakness of the travel plan approach. The main weakness is that the vast majority of employers and other institutions do not see solving transport problems as their responsibility. To date, rather than adopting an integrated management approach, employers have tended to treat transport matters as separate, self-contained issues, many of which are seen as largely outside an employer's control and thus not their responsibility. These issues have included:

- road congestion affecting delivery reliability and costs (as well as staff punctuality)
- congestion of onsite parking
- transport-related planning conditions required for site development
- changes to the tax treatment of transport benefits in the remuneration package (company cars, mileage allowances, etc.) and other transport-related human resource issues
- transport and company environmental policies (including environmental requirements of export markets and other supply chain pressures)
- transport effects upon brand image and public relations
- transport impacts upon company quality initiatives.

Figure 3.3 Traffic congestion affects delivery, reliability and staff punctuality

Furthermore, employer and institutional aspects of transport are now the subject of a major policy initiative via the UK Government's integrated transport policy. The rise in congestion and pollution is not bad just for the environment and society – it is bad for business as well. Congestion costs money and a variety of new measures are planned or under way that will have impacts upon employers, including:

- workplace parking charges (whereby in some cities there is a levy on each parking space on an employer's site)
- congestion charging (where motorists are charged to enter a city centre, as in London, Durham, Oslo and Singapore), also known as *road user charging* or *area licences*
- changes in company car taxation to favour 'greener' vehicles
- changes in general vehicle taxation to favour fuel efficiency and cleaner fuels for both cars and trucks
- measures to increase the choice and opportunity for travel by 'greener' forms of transport
- tax concessions for some employer-provided 'green' transport.

As in all good management practice, it is crucial not to treat these seemingly disparate issues in isolation: this leads to inefficient, ad hoc 'fire fighting' that is costly and ineffective. The key to a cost-effective and successful approach is to recognize that organizations are dealing with a series of challenges and opportunities stemming from key changes in the transport environment. A strategic, integrated business approach is needed, and this is where travel plans come into their own.

Nevertheless, companies generally consider travel plans only when some other pressing reason forces them to examine how their staff get to work

(Rye, 2002)[1]. In some cases this may be a crisis of onsite parking congestion, or a perception that a company's transport problems are harming its business image. To take an example very close to home, it could be embarrassing to The Open University, which teaches environmental management in its courses, if it could not show that it practises what it preaches by having an effective travel plan. However, in the majority of cases the most pressing reason arises when a company wishes to move into a particular area or expand its site, and the local authority forces it, through a condition of planning consent, to look at alternative ways in which employees or customers may travel. Such a planning condition is often called a 'Section 106 agreement', after the section of the 1990 Town and Country Planning Act that provides powers for councils to set such conditions. In Scotland this is called a 'Section 54 agreement', after the section in the comparable Act for Scotland.

For some public sector organizations, a travel plan is now required directly by government or as a condition of finance. When applied well, travel plans can cut car use by worthwhile amounts. The best employer travel plans in the UK have secured a reduction in car use of between 10% and 20%, while some feel that up to 30% is a possibility (Addison, 2002). In the USA, where mandatory travel plans have been in use, a 30% cut in car use has been achieved in several cases.

Incentive mechanisms for employers

In addition to providing information and guidance on why and how to introduce travel plans, there are effectively three other mechanisms to persuade companies to encourage their staff to commute in a 'greener' way. These are:

1 regulation

2 subsidies

3 the tax system

Regulation

Although in the UK a travel plan may be required in order to get planning consent for a site development, in some countries a whole regulatory framework governs how companies deal with their employees' commuting. We have already noted that air quality legislation in the USA included a mandatory requirement for larger employers to reduce driver-only car commuting to specified levels. This is now no longer required at the federal level (tax incentives are used instead), but several individual states have retained a regulatory requirement. In some places employers are required by law to subsidize their employees' public transport costs. Since 1983, in the Paris region of France, employers have been required to refund half the cost of the *Carte Orange* season ticket (Flowerdew, 1993). A similar scheme, the *Vale Transporte*, operates in Brazil.

[1] There are notable exceptions. For example, The Body Shop actively wants its brand to be identified as being 'environmentally aware', and so adopted a travel plan for many of its sites in the UK.

In Italy, the Government has begun pursuing a mandatory approach to travel plans. In 1998 the Environment Ministry mandated the Decree on Sustainable Mobility in Urban Areas. Organizations employing over 300 staff must designate a mobility manager to co-ordinate efforts to reduce employees' home–work trips through a site-specific 'mobility plan'. However, the impact of this measure is likely to be limited as no quantitative targets are set and there are no penalties for companies not complying (MOST, 2001).

Figure 3.4 The Paris Metro and (inset) the Carte Orange. Employers in Paris must subsidize the public transport tickets of their staff

Subsidies

An alternative to regulation is to use public subsidies as an inducement. In most cases subsidies are used to help organizations develop their travel plan programmes. For example, the Space Coast Area Transit Agency in Southern Florida, USA supports the Space Coast Commuter Assistance (SCCA) programme to help commuters use alternative transport to the private car. The agency assists businesses individually to develop programmes for reducing commuter trips, and makes no charge for its services (Litman, 2001). In Linz, Austria, the city council offers a free mobility consulting service to the 450 companies that have 50 employees or more (Schippani, 2002). In the UK, the government funds a free travel plan advisory service for employers and schools, providing 5–10 days of expert advice to each site. By early 2003 over 400 sites had been offered advice under this Site-Specific Advice scheme.

Direct cash subsidies are less common. However, in Montreal, Canada, cash subsidies have been offered to employers with more than 50 employees to develop travel plans (Coulliard, 2002). In Italy, the Environment Ministry has set aside €15.5 million over three years to finance up to 50% of the design and implementation costs of companies' mobility plans (MOST, 2001).

Figure 3.5 Companies in Ottawa can apply for a 15% discount on the cost of public transport tickets for their staff

Another form of subsidy occurs when public transport bodies (sometimes in partnership with the local authority) offer employers discounts for buying public transport passes in bulk. The Ecopass in Ottawa, Canada is one such scheme that is currently being piloted. Member companies obtain a 15% discount on annual passes, which are paid for through payroll deduction (OC Transpo, 2001).

Tax system

Tax incentives can be a powerful mechanism and have been used in the USA and the Netherlands to stimulate the development of travel plans by employers. For example, in the USA employers can subsidize the public transport fares or vanpool[2] costs of their staff by $100 a month (plus $155 for vanpool parking). If the employer does not subsidize public transport fares, individuals can buy tickets free of tax up to a specified allowance (IBI Group, 1999). Individual states in the USA sometimes have additional tax measures. One example is Oregon, where businesses can receive a 35% tax credit for their investments in trip-reduction activities, including telework equipment for their employees, vehicles for vanpooling, and bus passes (Litman, 2001).

Since 1999 the UK has introduced a series of tax concessions to support travel plans. However, the main reason for these concessions has been not to provide a positive tax incentive to travel plan measures, but to remove the tax disincentives. For example, until 1999 the benefit of employer-provided car parks was not taxable, whereas the same amount spent subsidising staff travel by cycle or bus would be taxed as 'income in kind'. A number of important travel plan measures have now been removed from

[2] A 'vanpool' is a minibus provided by a company and driven by one of their employees who picks up others in their 'pool' on the way into work.

the tax net, including private works buses, subsidies to improve the quality and coverage of bus services to an employer's site, and the provision of bicycles (Potter *et al.*, 2003). Bus fares can also be subsidized on routes to an employer's site, but subsidies for train, metro or tram fares remain taxable.

3.4 Transport impacts at hospitals

One employer that has come to view transport as very much its own problem is the National Health Service. As the NHS sees it, 'movement of ambulances, patients, visitors, staff, suppliers, contractors equals congestion, pollution and road traffic accidents' (NHS Estates, 2002a). Hospitals attract a great number of visitors and have a large number of employees. They also need to have good access for ambulances, to be accessible to people who have a variety of disabilities, and to receive all sorts of deliveries. At the same time hospitals should provide a tranquil atmosphere, where people can get well again. The locations of hospitals and other health institutions are often far from ideal in meeting all these needs. The NHS is the largest employer in Britain, with around a million staff. On top of that, it generates a million patient journeys each day. Hospitals are the largest generators of traffic outside peak hours and are estimated to account for up to 5% of all trips (Department of Transport, 1996). With around 70% of trips to and from hospitals being made by cars, the NHS contributed 2.1 billion car trips to Britain's roads in 2000 (Dublin Transportation Offico *ct al.*, 2001).

Figure 3.6 Hospitals are the largest generators of traffic outside peak hours

The general transport impacts of hospitals may be large, but transport is also an internal cost to hospitals, affecting their core functions. A survey in 1997/98 found that 38% of householders without a car found it difficult to get to a hospital and 16% said it was difficult to travel to a doctor (NHS Estates, 2001).

Other travel-related problems include:

- patients missing appointments
- visitors who are already under stress because a relative or friend is in hospital suffering added frustration and aggravation through inability to park or the cost of parking, especially for long-term hospital stays
- staff complaining that they cannot get to work and/or the cost of parking being too high
- local residents complaining that they cannot get to their own homes because hospital staff and visitors park in front
- local authorities complaining because the public highways become congested with parked cars, causing hindrance and inconvenience
- potential for bad congestion and poor traffic flows hindering ambulances getting patients to the hospital or Accident and Emergency (A&E) facility
- likelihood of increased road traffic accidents.

Figure 3.7 Many people find it difficult to get to hospitals (source: *Local Transport Today*/Nick Jeanes)

In addition, providing new parking spaces is very expensive. Each car parking space costs between £300 and £3000 per annum – money that might be better spent on improving the NHS.

The transport circumstances of hospitals were explicitly recognized by the Government in the 1998 White Paper *New Deal for Transport: Better for Everyone*. This stated that:

> [the Government is] particularly keen that hospitals are seen to be taking the lead in changing travel habits. By the very nature of their work, hospitals should be sending the right message to their communities on acting responsibly on health issues. We would like to see all hospitals producing green transport plans.
>
> Department of the Environment, Transport and the Regions, 1998, p. 141

Hospitals are at or near the top of the list of institutions required to contribute towards solving the transport crisis. Within the NHS itself, the desire to reduce the transport problems caused by healthcare sites is reiterated in a number of policy documents. For example, the *New Environmental Strategy for the National Health Service* (NHS Estates, 2002b) identified transport as one of five key areas where progress needs to be made; the others are procurement, energy, waste and water. Specifically, the document stated that NHS sites should have developed a 'healthy travel plan' by October 2002. This should be in line with the local authority transport strategy and identify the potential for reducing journeys and using smaller-engined, low-sulphur or LPG fuelled vehicles. The benefits would be increased fuel economy and lower tax, less pollution, better financial returns, and less stress from traffic jams. It is interesting to note that these benefits arise from a mixture of technical measures and behavioural change measures. Travel plans are not just about one approach or the other; the two are very much viewed as mutually reinforcing.

Another impetus for hospital travel plans came from the Department of Health's risk management process entitled 'Corporate Governance: Controls Assurance', which includes transport as one of its key criteria, while the National Service Framework on Coronary Heart Disease required that trusts develop travel plans by April 2002.

Sustainable Development in the NHS (NHS Estates, 2001) set out practical initiatives that NHS facilities could implement to minimize the impacts of transport at existing and new sites. In short, it suggested that hospitals need to:

- consider the influence of location on accessibility and transport impacts
- reduce the environmental impacts of people coming to NHS facilities, through promoting transport alternatives or car sharing for staff
- reduce the need for patients, visitors and staff to travel between sites, possibly through the provision of a wider range of health and other services by one local facility central to its community
- increase the amount of information and advice available without the need to visit NHS facilities (that is, over-the-telephone or internet advice, such as the NHS Direct service).

Travel plans can also minimize transport impacts before hospitals open. When the relocation of the Princess Margaret Hospital in Swindon was being carried out, suppliers were encouraged to send full lorries to the site rather than lorries containing just a few items. This required close co-operation between site workers and suppliers to ensure that large amounts of material did not have to be stored onsite. In addition, vehicles coming onto the site were regularly tested for emissions, a car-sharing and minibus transport plan was set up for site workers, and shared cars were allocated car-parking spaces in preferable areas (NHS Estates, 2001).

3.5 Case studies of hospital travel plans in practice

Although the requirement for hospitals to have travel plans in place has existed only since October 2002, some pioneers employed travel plans several years earlier. This section looks at two of these pioneering plans and at how they worked and evolved. The examples in the boxes below are abridged versions of case studies from the Department for Transport (2002).

Plymouth Hospitals NHS Trust

Derriford Hospital is located in the outer suburbs of Plymouth, some five miles from the city centre. Its travel plan was initiated following Plymouth City Council's refusal in 1995 to allow a major increase in car parking, and was formally part of a 'Section 106' planning agreement. For Derriford Hospital, the travel plan involved using money from car-parking charges to subsidize bus travel to the hospital. A good partnership with the bus operators and Plymouth City Council resulted in more buses entering the site and a major rise in staff bus use. Facilities for cyclists and car sharers were also provided.

BOX 3.4 Management of the Derriford Hospital travel plan

Introduction and reasons for the travel plan

In October 1995, the [Plymouth Hospital NHS] Trust submitted an outline planning application to Plymouth City Council for the initial element of a car parking strategy that envisaged the creation of 630 additional spaces on the Derriford Hospital site by January 1998. Plymouth City Council rejected the planning application on the grounds that it contravened central government land use and transportation policies as set out in Planning Policy Guidance Notes (PPGs), in particular PPG 13 …

Derriford Hospital was at this time in the process of planning significant developments at the site. Phase IV required extra parking spaces to allow for the transfer of some services from other health care sites elsewhere in Plymouth to Derriford. Approval from Plymouth City Council was eventually given for some extra parking spaces in return for the trust accepting a Section 106 planning agreement, which placed a ceiling on the number of spaces to be provided on the hospital site for patient, visitor and staff use. The Section 106 agreement also required the trust to devise and implement a Staff Commuter Strategy to discourage single occupancy car journeys. As part of the planning agreement the trust was required to make regular counts of the number of cars on site. This is undertaken each weekday at 10.30 am and at 2.30 pm and identifies the total number of empty spaces at peak times. Car parking figures are fed into a monthly board report so that, from the top down, the organisation is aware of the situation.

Co-ordination and management of the travel plan

[…]

There is clear management support for the travel plan. The Director of Facilities is supported by senior management including the Chief Executive. The travel plan is included in the Annual Plan and associated reports of the trust. There are examples of management leading by example: Facilities Directorate staff use public transport and car share. The Transport and Environment Manager is seen as the champion for the plan, as was a previous Deputy Chief Executive.

Funding

Staff car parking charges were introduced at Derriford to deter car use and to generate recurring income to cover the cost of the trust's travel plan. Funding for the travel plan comes from ring-fencing of the staff parking charges and is used principally to fund heavily discounted bus fares for staff as well as car park improvements. It is recognised that income obtained from patient and visitor car parking charges will have to rise in line with other similar hospitals, and that staff may also have to pay more for parking, if the cost of alternative modes of travel is to be met as demand for them rises.

Travel plan measures

These are comprised of:

- Improved cycle facilities
- Improved pedestrian facilities
- A car sharing scheme with parking charge exemption, priority spaces and guaranteed ride home
- Improved bus service provision and information
- Car parking charges
- Improved security on site.

Main emphasis: car sharing and improved bus services.

Travel plan effectiveness

The Derriford travel plan set the following targets, for achievement by January 2003. These have been derived by assessing the possible impact of a package of incentives to promote alternatives to solo driving.

- Reduce the number of patients and visitors who travel to the hospital in their own car during peak periods by 15% when compared to that of January 2000
- Ensure that patients and visitors are not required to search for longer than 10 minutes to find a vacant car parking space on the hospital site
- Encourage an increase in the number of direct bus routes serving the hospital during peak times by 15% compared to that of January 2000
- Reduce the staff parking space per employee ratio by 10% compared to that of January 2000.

[...]

[The change in the typical daily travel of staff before and after the travel plan was implemented showed a cut from 78% travelling in as a car driver in 1995 to 54% in 2001. In particular, bus use had more than doubled from its 1995 share of 8% of trips.]

Costs and benefits

The annual cost of funding the travel plan is approximately £150 000. In 2001 this was comprised of:

- Car sharing £200
- Bus measures £59 500 (+ £17 500 for national insurance)
- Publicity and promotion £3 000
- Cycling measures £15 000
- Staff time in managing the plan £16 500
- Bike/motorcycle interest free loan £6 480
- The remaining £31 820 is spent on maintenance, security, lighting, landscaping, pavements etc.

There have been some initial costs for setting up aspects of the travel plan in 1997–98 and in total it is calculated that this was £127 000.

The trust has also calculated that the upkeep and day to day operation of the site's car parking facilities, including demand management measures, costs the trust £445 000 per annum (i.e. £210 per space) at 2000 prices.

Annual running cost per member of staff (calculated as £150 000/4193 full time equivalent) is just under £36.

Support for bus and rail use

There is no local railway station. The main Plymouth station is located 5 miles away in the city centre.

Prior to the development of the travel plan in 1997–98 there were 22 bus services serving the site at peak hours. By 2001 this figure had risen to 44 buses. Derriford Hospital is consequently well served by public transport. The bus operators have restructured their services so that 80% of the existing routes serving the northern part of Plymouth provide direct and frequent access to the hospital. In collaboration with the city council and the bus companies, the trust has joint funded and produced a Travel to Derriford leaflet with bus timetables.

In 1997 there were two bus stops on site. This has risen to five with three bus shelters and set up and set down points. The hospital's bus lay-by has been trebled in size in order to cope with the higher volumes of bus traffic [and the] trust has ... agreed to the creation of a purpose designed bus station on the hospital site. This is being funded through revenue from Plymouth City Council's local transport plan.

There is an array of discount subsidised bus passes available. The original was the Derriford Travel Pass available to staff handing back their car parking permits. This involved a half price ticket, which was 40% subsidised by the trust and 10% by the public transport operator. From April 2000 the trust has offered a four-month trial free bus pass to staff for handing back a car-parking permit. At the end of the four months staff continuing to use the bus can get a 65% reduction on a bus pass for 12 months of which 55% is trust subsidised and 10% from the public transport operator. Further bus passes have a 50% discount. By 2000, 443 discounted tickets paid for by the trust amounted to £80 130.

For other staff, there is a Green Zone Bus Pass, introduced from April 2000, comprising of five zones. This discounted ticket has been negotiated with other local employers and the city council. The Green Zone Bus Pass gives a 25% reduction on the standard single bus journey ticket. For journeys within 5 miles a monthly ticket costs £29.25, for 5–10 miles £36.00, for 10–15 miles £42.50 a month, 15–20 miles £48.00 and 20–25 miles £53.00. The tickets are valid for bus services provided by both main operators in the city, Plymouth City Council, and First Western National. Other ticket offers include 10 journeys for the price of 12 and again these are valid with both the main bus service providers.

The trust has funded discounts on two routes to the hospital operating through areas of poor health. In addition, it has encouraged visiting between 6–8 pm through cheaper parking rates. The latter has resulted in a 25% increase in evening visiting since 1999.

Support for cycling

Access to the site is reasonably good by bicycle. There have been both off and on-site improvements for cyclists since 1997. Off site measures have been developed independently by Plymouth City Council. On site facilities include shortened road humps so that cyclists can avoid these. There were existing showers and changing rooms which can be used by cyclists and extra lockers were introduced in 1997–98. In 1997 there were no bicycle parking facilities but by October 2001 there were 100 spaces.

From 1998 the trust has offered staff a £500 three-year interest-free loan for the purchase of a bicycle. Cycle training is offered to staff but there have been no demands for this. The trust has produced a one off newsletter in June 2000, Pedal Power.

Support for walking

Accessibility of the site by foot is described as 'medium'. The trust has pressed the city council to make improvements to off site pedestrian facilities. Regarding on site facilities, the trust has completed development of a pavement network. In 1998–99 five zebra crossings were installed (and one removed). The trust has also funded improved lighting.

As with cyclists, pedestrians have access to showers, changing facilities and lockers. There is a contracted security patrol service operating across the site.

Support for car sharing

The trust has operated a computerised matching service since 1997–98. In 2000 the names of 640 potential car sharers were contained on the computer database. Car sharers are exempt from car parking charges and have priority parking spaces closest to the hospital buildings. There are 130 car parking spaces reserved for car sharers ... Since April 2000 there is a guaranteed [taxi] ride home should the planned ride home not be available due to unforeseen circumstances.

Car park management

Some 54% of staff have parking permits and these tend to be 'front-line' staff involved in patient care, disabled staff and those required by contract to have use of a car. Claims for permits on the grounds of travel during the course of work are checked against mileage claims and evidence of need. For those with occasional need to bring a car on site (for example, when bringing in heavy equipment) there are one-day permits.

In 1997–98, a 20p a day charge was made for car parking on site on weekdays. This was increased in 1999 to 50p a day. Staff can choose to have charges taken from their salary or pay in coins on each occasion. Annual charges are made on the basis of 252 working days minus four weeks leave and two weeks for sick leave. Weekend parking has remained free. Night staff, weekend staff, disabled staff, volunteers, car sharers and tenants of the site's residential accommodation are currently permitted to park their cars on site free of charge. Parking permits are not required out of hours (though medical shifts begin or end at times when permits are required).

There is a financial incentive for staff to return their parking permits: this cash-out scheme gives staff who drive to work on at least three days a week, £250 for surrender of a permit. This incentive has been on offer since June 2000 and seven permits had been surrendered by October 2001.

The trust operates an appeals procedure in which the Director of Facilities is the final arbiter. Any changes in charges or benefits arising from the travel plan have to be approved by a Joint Staff Committee.

Other strategies

Since 1997–98 all applicants for posts at Derriford Hospital have received an applicants' pack, which contains information about the travel plan and the parking constraints, which might help with location choice for new employees who are considering moving into the area. There are moves to promote more flexible working to help reduce travel demands on the site. The Facilities Directorate has negotiated with departments in offering up to £450 to help existing car driving staff to work from home on some days, although this has not yet proved attractive to departments.

The trust does provide personalised travel planning advice to staff on request regarding public transport and car sharing.

Communications

Since 1997 the trust has communicated with staff on a continual basis about travel plan developments through a variety of media. The mechanisms for this communication have been partly through posters and newsletters.

Between 1997 and 2000 the trust produced the Derriford Newsletter to inform staff about the travel plan. Since June 2000 specific mode newsletters have been produced. There is discussion with staff consultation bodies through a Joint Staff Committee, which meets quarterly or when needed. The group is comprised of five union representatives and two or three managers.

The city council has provided some posters and the bus companies bus travel literature including bus maps. There is an annual bus road show and in 2001 a road safety show during road safety week run by the city council road safety team. This was particularly popular in covering issues such as child car seats and injuries caused to pedestrians and cyclists on the roads.

The trust works closely with the city council and acts as a lead on travel planning for other employers. The liaison officer for the trust at Plymouth City Council is the trust's previous Transport and Environment Manager, which makes collaboration and understanding much easier than it might otherwise be. There are quarterly meetings with the city council and the public transport operators at which, for part of the meeting, trust staff can ask questions about services.

The trust has made use of Geographical Information Systems to target staff living close to bus routes or where there is potential for car sharing. Letters are then sent to the specific staff members about the options available to them.

Views of those managing and implementing the plan

According to the Transport and Environment Manager, the trust suffered from a lack of available experience when it started its travel plan. There was no advice available at the time. The trust did receive original advice in 1996 from a consultant and has since maintained good links with Transport 2000, contacted similar organisations, looked at student materials and gathered what information it could. In 1996, however, no public sector organisation the trust knew of had been refused planning permission and so this was unexpected. Yet the refusal marked a turning point in that the trust was resigned to developing a travel plan.

There were some early backlashes, including a junior doctors' motion of no confidence in the hospital management, but this settled down after a while and by 1998 the local paper started to respond positively towards the travel plan. Nonetheless, to implement a successful strategy it is important to be able to communicate and to have charismatic managers, and to be able to manage behavioural change. This requires being 'thick skinned' and having motivation and ongoing support from within and outside the organisation. It is also important to understand that a travel plan is a living document and has to be regularly updated.

The two greatest successes have been the increase in bus use and in car sharing for which there have been high levels of support from staff. The results of these have been to reduce congestion on arterial routes into the hospital, support public transport and promote a choice of modes.

In developing the travel plan it has been important to have a good relationship with the city council, especially the person working on travel plans. Plymouth City Council has been very supportive, and has underwritten some bus routes, and put in their own funds. It has similarly been important to have good working relationships with the bus operators and to make the business case for services.

Department for Transport, 2002, pp. 108–13

(a)

(b)

(c)

Figure 3.8 The sorts of measures used at Derriford have been applied by many other employers: (a) special car sharer spaces at Boots, Beeston; (b) company-supported bus service at Orange, Bristol; (c) modern covered cycle and motorcycle parking at

Nottingham City Hospital NHS Trust

Nottingham City Hospital NHS Trust is another example of a more established travel plan. The Trust's activities are centred on a large edge-of-town hospital site about six miles from Nottingham city centre. In 2002 it was estimated that 12 000 vehicles a day entered the site; there were 1200 parking spaces for staff and 450 for patients and visitors; the hospital employed 5200 full- and part-time staff; it had 250 000 outpatient appointments and treated 75 000 inpatient and day cases. Like Plymouth, the Trust entered a 'Section 106' planning agreement in 1997 to develop a travel plan. Ring-fenced funding from car-parking charges financed better pedestrian and cycling provision and measures to enhance public transport. Travel surveys revealed that between 1997 and 2000 driver-only ('solo') car use declined from 72% to 55%, while car sharing rose from 2% to 11%, and bus use increased from 11% to 19%.

BOX 3.5 Management of the Nottingham City Hospital travel plan

Introduction and reasons for the travel plan

In 1996 the [Nottingham City Hospital NHS] Trust provided free parking and had an unknown number of vehicles entering the site. There was little security on site and between 70–80 vehicles per month were subjected to car crime. No public transport entered the site, there was little understanding of pedestrian requirements, and there was one dilapidated cycle shed. Unrestrained car use had resulted in gridlock on site at peak times, parking chaos, and little faith in security. Additionally, the trust was entering into a Section 106 planning agreement for the construction of new buildings on the site and needed to have a co-ordinated approach to travel planning.

It is seen as essential for the trust to have a coherent travel plan in order that support facilities such as car parking are adequate to enhance the 'patient experience'. Therefore the trust has the following objectives:

- to develop a strategy for the future (2001–2006)
- a menu based approach, whereby it allows the trust board to tailor the proposals to best meet service requirements
- to provide sustainable alternatives of transportation to and from the hospital
- to ensure patients and visitors receive a quality service.

Coordination and management of the travel plan

The trust produced a first travel plan in 1997. This involved negotiations with staff and their representatives and feedback was that any travel plan funds generated from parking revenue had to be ring-fenced for transport improvements.

There is management support for the travel plan and from autumn 1996 it was incorporated into the corporate strategy for the trust. There are examples of management leading by example through the returning of managers' parking permits on account of the high frequency and low cost of bus services to the city centre. The Chief Executive has also given his personal approval to the travel plan and there is endorsement by the trust board.

Funding

Funding for the travel plan comes from ring-fencing of the parking charges. The annual cost of funding the travel plan is approximately £144 000. This is comprised of £100 000 capital to spend from car parking revenues once payments have been made for park and ride, parking management (contracted to outside service) including parking wardens, and CCTV cameras. A sum of £15 000 is spent on cycling each year and approximately £29 000 is staff costs.

Travel plan measures

These are comprised of:

- Improved cycle facilities
- A car sharing scheme
- Improved public transport provision and information
- Car parking charges
- Improved security on site
- Park and ride.

Main emphasis: car parking charges and buses on site.

Travel plan effectiveness

[A comparison of 1997 and 2000 staff travel surveys indicated that] solo car driving had reduced significantly and that bus use had increased by 73%.

Staff: Main mode of travel to/from work

Mode	November 1977 %	November 2000 %
Pedal cycle	5	4
Car (drive alone)	72	55
Car sharer	2	11
Bus	11	19
Train	0	1
Walk	8	9
Other	2	1
	100	100

[...]

Costs and benefits

As noted above, the trust spends about £144 000 a year on the travel plan. In 2001 this is comprised of:

- Car sharing £2 000
- Bus measures £8 000
- Publicity and promotion £1 500
- Cycling measures £15 000
- Staff time in managing the plan £29 000
- Walking measures £60 000
- Signage and maps £28 500

There have been some initial costs including £112 000 for the installation of CCTV cameras which was capitalised over the length of the contract. Each year the contracted parking service costs £180 000.

The annual running cost per full time equivalent employee is £41/employee (figure excludes revenue from parking).

The main benefits of the travel plan have been that it has given staff, patients and visitors a range of sustainable transport alternatives to solo car driving, together with informed choice about these options. It has been critical to put in place the 'carrots' rather than to start with 'sticks' so getting in infrastructure has been important. The most successful aspects of the travel plan have been increases in bus use and maintenance of cycle use. In particular, the introduction of buses on the site was critical in bringing about increases in bus use.

Support for bus and rail use

There is no local railway station. The main Nottingham station is located six miles away near the city centre although buses from the railway station enter the hospital site every 30 minutes.

Prior to the development of the travel plan in 1997 no public buses entered the site as services only stopped at the periphery (which is more than 400 m from the building entrances). By 2001, there were services entering the site every 15 minutes during weekdays between 7 am and 6 pm, (including between 8 am and 9 am). These services are operated by Nottingham City Transport who have funded bus shelters, a new fleet of low floor buses, and a travel map of their routes serving the hospital site ... There are also more services that pass the hospital periphery.

There are currently no specific discounts on bus service fares generally available to employees, but this is being pursued. The cost, however, of a single ticket to the city centre at 70p makes the bus journey attractive to staff, especially as Nottingham City Council (highway authority since 1998) has introduced more bus lanes around the district. Nottingham City Transport provides a 28 day bus pass for £28.00 (£3 for initial provision of identity card) which provides unlimited travel. There is information about bus services on the hospital web site and also on the intranet for staff which have hyperlinks to Nottingham City Transport and Trent Barton Buses, the main bus service providers in Nottingham.

The trust operates a park and ride service within the site, running every 15 minutes using three minibuses, funded out of car park revenue. Two of the vehicles have been donated by the Women's Royal Voluntary Service and the hospital's League of Friends.

Support for cycling

Access to the site is reasonably good by bicycle. There have been both off and on-site improvements for cyclists since 1997. Off-site improvements include routing part of the Nottingham cycle network past the front of the hospital site. On the site, there were existing showers and changing rooms which could be used by cyclists and these are to be upgraded in 2002. In autumn 2001 there were 450 cycle stands on site ... There are also three secure cycle compounds that can hold 90 cycles. These are remotely patrolled through CCTV cameras installed in 1998.

A Bicycle Users Group was established in 1997 although this has evolved into the alternative transport group within the hospital which focuses on all alternative modes to solo car use. The trust takes part in a range of cycling promotion events, including the annual Bike Week in June. It has a fleet of 12 bicycles for staff use and the trust pays 11p a mile for travel during the course of work. The bicycles are maintained by Raleigh (bicycle manufacturer located in Nottingham) and staff have access to lights, locks, baby seats, helmets and car racks. This is a popular service. Staff can take advantage of a 20% reduction on cycle equipment from Raleigh and a 12% reduction on the cost of a new bicycle. There is an interest free loan available for bicycle purchase.

[...]

Support for walking

Accessibility of the site by foot is described as 'medium'. The trust has employed consultants to advise on improvements for pedestrians in recognition that several hundred of its staff walk to work each day. A 15 mph speed limit has been introduced on the site with some cycle-friendly traffic calming measures, dropped kerbs, and new pedestrian zebra crossing installed at a cost of over £100 000. Street lighting has been upgraded and some new paths constructed.

There is also a programme of pedestrian signing being introduced or upgraded. This has arisen from an audit of the site by consultants and the development of a Pedestrian Signing Strategy in 2000.

Support for car sharing

The trust has operated a computerised matching service since June 2001. Staff can access this via the intranet and self-match. There are plans to exempt car sharers from parking charges in the revised travel plan for 2001–2006 and priority parking spaces nearer to buildings.

Car park management

There is an annual car parking charge for staff of £55.00. Each staff member can apply for a permit allowing access to the site. Staff car parking charges were introduced in 1997 at £50 a year and raised in 1999. There is a window sticker permit. As of September 2001 all students are banned from bringing cars on to the site. Currently some staff with peripatetic work patterns, such as some surgeons and community nurses who work off-site are guaranteed a parking space. In 1997 this was 3.8% of staff and the percentage has reduced to 3.2% in 2001 against a background of stable staff numbers.

There are currently 1200 car parking spaces dedicated for staff use, with 4000 'live' car parking permits issued. This results in as many as 200 staff cars parked on an unofficial basis each day.

Because of substantive improvements in car park security arising from travel plan measures the trust has received ten car parking awards since 1997. There is a trust security working group which in 2001 has been evaluating how other large organisations manage their security issues in order that an integrated system can be developed at the hospital site.

[...]

Communications

Since 1997 the trust has communicated with staff on a continual basis about travel plan developments. The mechanisms for this communication have been through the hospital newsletter City Post, road shows about the travel plan, articles for the hospital notice board, information included in pay packets, and emails. The trust alternative transport group has been exploring the idea of a logo to give the travel plan a unique identity.

Views of those managing and implementing the plan

According to the Environmental Services Manager, it is important to expect some bad publicity and to have a 'thick skin'. It is however, important to get communications right and so to keep the media well informed, including the internal public relations staff and the local media who will always be quick to publicise perceived opposition. At the trust, the staff and their representative wanted to see clear evidence that money raised through parking was being reinvested in transport security measures. Continual liaison with staff groups and use of internal communications media was, therefore, important. In addition, there must be support from the highest levels of management for the travel plan.

The trust has had some very good publicity from its work on the travel plan, nationally and internationally. It has also developed a very good working relationship with Nottingham City Council since 1997, with whom it had little contact prior to this time. Similarly it has developed good working relationships with the local bus companies.

In terms of plans for the future, a key objective is the implementation of a new travel plan with restrictions on parking according to distance from home to hospital. Subject to further negotiations with staff, those new staff living less than 800 metres [away] will be barred from driving to work unless they have special justifiable reasons for the use of a car. The cost of permits is to be structured so that those living closest to the hospital will pay most for a car parking permit. Staff working shifts or on rotas will be given higher priority for permits as well as those who car share.

The trust also wishes to increase bus services further and to develop a transport hub within the site. This would provide:

- a focal point for public transport
- travel information
- toilet/baby changing facilities
- facilities to meet special needs
- snacks and beverages
- travel ticket issue.

These facilities would enhance in essence the government's initiatives (The NHS Plan) of providing patients with focal points for information on a personal level.

The new travel plan will have a range of targets to be achieved. These are set out below:

Criteria	From:	To:	Date
Monitor air quality and vehicle count	Ongoing	Ongoing	May 2001
Increase disabled car parking spaces	90	150	April 2003
Increase bus use	19%	21%	April 2004
Increase cycling use	4%	5%	April 2003
Reduce single car occupancy	55%	50%	April 2005
Establish car share database	May 2001	May 2004	May 2004
Develop car parking facilities	December 2001	December 2002	December 2002
Develop travel hub/Light Rapid Transit	May 2001	May 2002	May 2002
Reduce day time deliveries by fuelled vehicles	May 2001	May 2002	May 2002
Increase patient parking facilities	480	600	May 2003
Undertake travel survey	Bi-annually		May 2002 2004 & 2006

Department for Transport, 2002, pp. 78–82

Figure 3.9 Covered cycle parking at Addenbrooke's Hospital, Cambridge. A travel plan has been introduced here also

Overall, the performance of travel plans in hospitals has shown that, once transport management is accepted as a legitimate function of an institution, it can be applied effectively and efficiently. As noted above, at Nottingham City Hospital driver-only car use dropped from 72% to 55%, with rises in bus use and car sharing particularly noticeable. At Plymouth, over a comparable period of time, the drop in car use was from 78% to 54% – an achievement remarkably close to Nottingham's, using similar measures. Other hospitals have also achieved comparable results from their travel plans. At Addenbrooke's NHS Trust in Cambridge the drop in car use between 1993 and 1999 was from 74% to 60%. Here, as well as bus use rising from 4% to 12%, cycle use rose from an already healthy 17% to 21% (Department for Transport, 2002).

3.6 Rising from the 'bed of nails'

The case studies above show that the use of mobility management measures allows institutions to achieve quick results in cutting energy use and emissions. But these measures do need to be tailored to the institution's needs and often require a regulatory kick-start (in these cases a planning requirement). By its very nature, transport policy is frequently subject to disagreement and controversy. The active involvement of institutions such as hospitals in developing mobility management strategies that meet their own needs could mark an important step towards a more consensual partnership approach. This may be one way to rise from the 'bed of nails'.

The next chapter examines travel plans more generally and how they are starting to be used by a variety of private and public sector organizations.

References

Addison, L. (2002) 'Using the planning process to secure sustainable transport', *Association of European Transport Conference*, Homerton College, Cambridge, 9–11 September, ECOMM (on CD-ROM).

Coulliard, L. (2002) 'Mobility management in the Montréal region: partnership strategies and transportation management areas', Economic Community Workshop, *Proceedings for the European Conference on Mobility Management*, 15–17 May, Ghent.

Department of the Environment, Transport and the Regions (1998) *New Deal for Transport: Better for Everyone*, (White Paper), London, The Stationery Office.

Department of Transport (1996) *Transport Secretary Urges Hospitals to Reduce Reliance on the Car*, press release No. 291/96, 17 September, DoT.

Department for Transport (2002) *Making Travel Plans Work: Case Study Summaries*, London, The Stationery Office.

Department for Transport, Local Government and the Regions (2001) *Evaluation of Government Departments' Travel Plans*, Report for the DTLR, London, April (unpublished).

Dublin Transportation Office, Kirklees Metropolitan Council and the Irish Energy Centre (2001) *Impacts Calculator, The Route to Sustainable Commuting: An Employers' Guide to Mobility Management Plans*, Way to Go Research Project, European Commission SAVE II Programme, Brussels, European Commission.

Energy Efficiency Best Practice Programme (2001) *A Travel Plan Resource Pack for Employers,* Energy Efficiency Best Practice Programme, London, The Stationery Office.

Flowerdew, A. D. J. (1993) *Urban Traffic Congestion in Europe: Road Pricing and Public Transport Finance*, London, Economist Intelligence Unit.

Freund, P. and Martin, G. (1993) *The Ecology of the Automobile*, Montreal, Black Rose Books.

IBI Group (1999) *Tax Exempt Status for Employer Provided Transit Benefits*, Final Report to the Canadian National Climate Change Process, Transportation Issue Table, chaired by Transport Canada, Ottawa, IBI Group.

Litman, T. (2001) 'Commute trip reduction (CTR): programs that encourage employees to use efficient commute options', *TDM Encyclopedia*, Victoria, Canada, Victoria Transport Policy Institute, http://www.vtpi.org [accessed 15 August 2001]

MOST (2001) 'A ride around Rome: implementation of the Italian MM decree, mobility centres and mobility consulting', *MOST News*, No.3, December, p. 3. Also available at http://mo.st/public/reports/most_news3.zip [accessed 10 June 2003]

NHS Estates (2001) *Sustainable Development in the NHS*, London, The Stationery Office.

NHS Estates (2002a) *Sustainable Development: Transport*, http://www.nhsestates.gov.uk/sustainable_development/index.asp [accessed 23 May 2003]

NHS Estates (2002b) *New Environmental Strategy for the National Health Service*, London, The Stationery Office.

OC Transpo (2001) *Ecopass: OC Transpo's Transit Pass Payroll Deduction Program*, Ottawa, Canada, http://www.octranspo.com [accessed 23 May 2003]

Potter, S., Enoch, M. and Rye, T. (2003, forthcoming) 'Economic instruments and traffic restraint', Chapter 16 in Hine, J. and Preston, J. (eds) *Integrated Futures and Transport Choices: UK Transport Policy beyond the 1998 White Paper and Transport Acts*, Aldershot, Ashgate.

Rye, T. (2002) 'Travel plans: do they work?', *Transport Policy*, Vol. 9, No. 4, October, pp. 287–98.

Schippani, R. (2002) 'Mobility consulting for companies through the City of Linz', Economic Community Workshop, *Proceedings for the European Conference on Mobility Management*, 15–17 May, Ghent, ECOMM (on CD-ROM).

Steer Davies Gleave (n.d.) *Travel Blending*, London, Steer Davies Gleave.

Chapter 4

Travel Planning

by Stephen Potter and Marcus Enoch

4.1 Introduction

Chapter 3 introduced the general concept of travel plans and looked at some examples of their use by hospitals. In this chapter, the process of developing a travel plan is explored in more detail, together with further evidence of the impacts on transport behaviour that can be achieved. The chapter will also examine how the concept of mobility management can be applied to freight movements as well as to personal travel. The chapter ends, in Section 4.7, by drawing together the various institutional responses considered in Chapters 2 to 4 and relating these to the transport challenge posed in Chapter 1.

4.2 Institutions and mobility management

Travel plans are a relatively new departure by UK employers, and have often originated not through strategic or corporate planning, but from ad hoc initiatives in response to a particular need. For example, many initiatives that may now be labelled as a travel plan were a response to a planning requirement or a parking problem. However, 'best practice' organizations – those against which others benchmark – have adopted a more comprehensive approach and discovered that travel plans can be justified in terms of improved efficiency and can yield cost saving benefits across the organization. The best organizations have integrated their travel plans with actions to clean vehicle emissions and cut energy use, such as introducing cleaner-fuelled vehicles into their company car or delivery fleets. Thus some public and private sector employers have repositioned their travel plans at the strategic level. This strategic approach has led to partnerships with local authorities, transport operators, other businesses and national government that have helped tackle bigger transport issues beyond the control of an individual employer. However, such best practice examples are rare. In general, travel plans are, at most, only at the fringes of an organization's agenda. Institutions are at the early stages of accepting that management of staff, customer and visitor travel is their responsibility. There are thus major cultural and institutional barriers to travel plans, which will take some time to overcome.

Because travel plans are an emerging transport policy response, and depend so much on institutions themselves accepting that they have a role to play, this chapter will look in some detail at the process of planning and implementing a travel plan.

4.3 Identifying the benefits of travel plans

As noted in Chapter 3, a travel plan can incorporate a range of transport-related initiatives aimed at addressing different transport aspects, including:

- commuter journeys
- business travel undertaken by employees
- customers and visitors
- deliveries
- fleet vehicles operating as part of the organization.

One problem with tackling the transport impacts of an organization is that any policy initiative tends to be viewed as an externally imposed regulation. In such a case, the usual response is to find the cheapest way to achieve compliance and leave it at that. It is thus not surprising that, when travel plans are introduced as a result of a planning condition, or a strategic NHS requirement upon existing hospitals, they come to be viewed in this way.

This in itself is a serious problem, since the idea that there can be *benefits* in managing the travel of staff, customers and visitors is rarely acknowledged. This is probably the most fundamental barrier facing the development of travel plans in the UK, and indeed in other countries where similar programmes have been introduced. Benefits include reducing the cost of employee business mileage, car park provision and maintenance. Research into the costs and benefits of good travel plans (Department for Transport, 2002a) showed an average cost for a travel plan as £47 per annum per employee, compared with a cost of providing car parking of at least £500 a year. Several companies, after developing travel plans, have recognized that it is something they should have done anyway as it made good business sense. The institutional response of building more and more car parks had never been previously challenged, yet it was not cost-effective. Thus, although the question 'could we build the car parks more cheaply?' may have been raised, the question 'would it be cheaper to accommodate these travel demands in ways other than building car parks?' never arose.

Why is this? Entire estates departments of major employers have built up an expertise in building car parks and access roads; they do not have the skills to, for example, organize the marketing of bus services or managing a car share database. Travel plans are not part of their normal culture or expertise, and require new ways of doing things. This is a general feature of the 'intelligent consumption' approach to energy efficiency and conservation. Added to this, the financial authorization of car parking is an accepted cost and, in principle, raises no problems. It is part of what an estates department should do. The activities of a travel plan, in contrast, have less institutional acceptance. Whereas a bid for money to build car parking may be rapidly approved, that for a cross-departmental programme of travel plan measures is likely to be more difficult to set up, be subject to delay and require greater justification. It may even be thrown out because there is no appropriate budget to finance it, or because financial rules do not permit a travel plan as an alternative to car parking. For example, a number of years ago The Open University ran buses to bring its staff into its Walton Hall headquarters. However, it was forced to stop doing this by the then Department of Education, which said that running buses was not a function the OU should undertake – whereas the OU's bid for a consequent major expansion of our car parks (costing more than the buses) was accepted without difficulty. Times and policies have changed (the OU does now financially support bus services to its site), but in other private and public sector organizations similar funding rules can still be in place.

An organization's response to an approach involving technical improvements to vehicles is quite different. An organization that runs a fleet of vehicles or provides company cars to its staff will be reasonably amenable to measures, such as those examined in Chapter 2, to introduce cleaner technologies and fuels. It will already have in place people, fleet management and financial systems to buy and run vehicles. 'Green' policy initiatives for vehicle fleets thus slot in with little or no difficulty. The organization can easily check on the costs of different vehicles and fuels, and respond well to financial incentives such as subsidies and tax incentives.

The benefits of a travel plan can extend well beyond cutting car-parking costs. It can have indirect benefits for quality issues such as the punctuality and health of staff as well as broadening the management of a company's car, van and freight vehicle fleet to encompass strategic transport issues. The key point of a travel plan is that it integrates the various transport provisions of an organization to ensure they complement each other and yield collective benefits.

In a more strategic sense, a travel plan can be part of a long-term environmental strategy. It can be a distinctive part of demonstrating to customers, shareholders and government the organization's environmental responsibility. Not least, a travel plan is a good insurance policy against

Figure 4.1 The staff commuter centre at Park Royal business park in London. Here the travel plan is marketed as a service to employees on the site

tightening planning and regulatory controls, as transport impacts begin to climb up the policy agenda of both local and national government. This is an issue that will not go away, but rather is set to increase in importance (as was demonstrated in Chapter 1). So, quite aside from the community and environmental benefits of a travel plan, there are private gains as well. However, these gains to an organization tend to be spread across several departments and are not immediately evident or specifically identifiable. This compounds the problems of image and culture involved in justifying a travel plan within an organization.

4.4 Developing a travel plan

When an organization is developing a travel plan, it is important to take into account the specific needs and circumstances of each site. In particular, travel plan development depends on such factors as organization size, location, the nature of the business (which influences the amount of business travel, number of visitors, number of deliveries, etc.), the reasons why a travel plan is being developed, staff attitudes towards different measures, and the resources available. For example, a travel plan for a large retailer's depot, which has a considerable amount of heavy goods vehicle (HGV) traffic and where the workforce all live locally, will be different from a travel plan for the national headquarters of a high-tech electronics company, where commuting is the main transport impact and most employees live further away. Finally, a travel plan is a continuous programme, and must be resourced, maintained and reconsidered periodically as the requirements of an organization change.

There are a number of stages that organizations need to go through when introducing a travel plan. The box that follows is taken from the Cheshire County Council Travelwise website. To follow the terminology used in the rest of this chapter, however, we have replaced the term *commuter plan* with *travel plan*.

BOX 4.1 Travel plans in Cheshire: steps to success

Step 1 – Identify the problems and make the case for action.

Step 2 – Secure commitment and allocate resources.

Step 3 – Raise awareness and build consensus with employees.

Step 4 – Gather data.

Step 5 – Review and evaluate alternatives to the car.

Step 6 – Agree the strategy and set targets.

Step 7 – Make it happen and maintain momentum.

Depending on a company or site's particular needs and objectives, the travel plan may form part of a wider company transport plan. This can cover a wide range of issues including commuting, business travel, [visitor and customer travel,] fleet management [including alternative fuels], deliveries and other commercial activity.

Step 1 – Identifying the problems

As already highlighted, there are many factors which can influence the decision to adopt such a strategy. These include:

- Concern about the impact of traffic congestion ...
- Pressures of on-site car parking demand ...
- Expansion plans leading to significant on-site development ...
- Environmental considerations ...
- Leading by example ...
- Conditions of planning consent ...

[...]

Step 2 – Secure commitment and allocate resources

It is essential that all Directors and Senior Managers understand and support the aims and objectives of the travel plan and are prepared to lead by example. This is vital to win over staff and Trade Union support and co-operation. [It is notable that this was one of the factors considered to be of importance in the hospital examples in the previous chapter.]

- Set a challenge to your senior managers to attempt to reach the workplace without using their cars.
- At least one senior manager should sit on, or chair, a steering group responsible for guiding the project's development.
- Appoint a dedicated travel plan / staff travel co-ordinator to lead the project's development.
- Secure funding to support the successful development of your travel plan.

Successfully introducing a travel plan can be greatly assisted by a variety of measures that can help send the right messages to all employees about the company's commitment to their particular programme of activity. Suggestions include:

- Review your existing company culture. Does this act to encourage car use? Can steps be taken to promote a more sustainable approach to travel.
- Consider your senior managers and directors giving up reserved parking spaces and pledging to use alternatives to their own cars whenever appropriate, to lead by example.
- Review your car park management and entitlement to parking permits. It may be suitable to revise allocation with priority given to work related needs.
- Ensure that all maps and guides showing your company's location for visitors and clients include details of how to reach the site by public transport and cycle.
- Ensure that your company includes staff travel information in new starter induction packs. This may include local public transport timetables or a registration form for a car sharing scheme.

Step 3 – Raising awareness

The key to success is staff 'ownership' and involvement. Your employees will need to be informed regularly about what the travel plan is trying to achieve, how it is doing this and, most importantly, what benefits will be gained by individual employees as well as the company as a whole.

Step 4 – Gathering data

Before decisions are made on what to include in your travel plan it is vital that you gather data on existing travel habits and alternatives.

The survey will act to review 3 key issues:

- Where people live.
- How they currently travel to work.
- Their willingness to use alternative types of transport instead of their cars.

This may require quite detailed questioning to understand factors that influence existing travel patterns and the necessary measures which may encourage people to use alternative types of transport.

It is also useful to build up a company profile. This will include details of workforce size, hours worked, number of car park spaces provided and the cost of this, existing measures to encourage alternatives to the car and consideration of future expansion plans.

Step 5 – Review and evaluate alternatives to the car

For many companies and their employees introducing a travel plan will mark a major break in prevailing company culture and car dependency. The preparation of the travel plan must take this into account. [This will be considered later in this section.] Your staff need to be satisfied that the proposals are not anti-car (many staff will feel that they have no real alternatives to using their car), neither should it impose the impossible or unworkable in its recommendations. Instead, it should build on information gained from the staff travel survey, particularly details about employees' willingness to switch [travel] modes and the measures required to bring this about.

[...]

Step 6 – Agree the strategy and set targets

In setting targets, the overall aim should be to seek:

- A reduction in single occupancy vehicles accessing your site.
- To increase the use of alternative modes of transport to the car.
- To set targets. The Government's Advisory Committee on Business and the Environment recommends that companies set a 10% target to reduce the number of employees commuting to work as the single occupant of a car.
- To be realistic, success will not come overnight. However, it is not unrealistic to consider a reduction in car use of at least 10% and maybe as high as 30% over a three year period, depending on location.
- To have regular reviews will greatly assist the success of the plan.

Step 7 – Make it happen and maintaining momentum

Get the formal launch of your travel plan right. This will pave the way for its success ...

This launch could include a challenge to help create staff ownership of the project.

The success of the travel plan depends very much on the level of your commitment, the resources available and the perceived attractiveness of the alternatives.

Appointing a dedicated staff travel co-ordinator will prove a crucial move in bringing these activities together and helping the overall development of the travel plan.

Cheshire County Council Travelwise, 2002

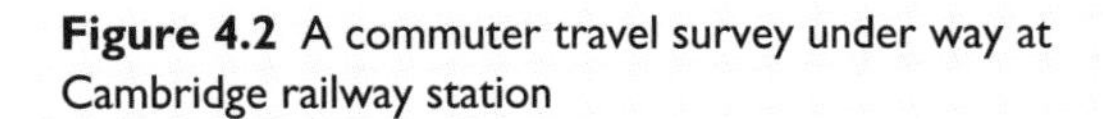

Figure 4.2 A commuter travel survey under way at Cambridge railway station

Travel plans and institutional politics

Steps 2 and 3 above raise the issue that travel plans, and the changes they involve, may not be welcomed by some employees and users within an organization, even if it is for the benefit of all. This crucial issue is addressed in the Department for Transport's *Travel Plan Resource Pack for Employers*, from which the following edited excerpt is taken.

BOX 4.2 Working with human resources staff and trade unions

A successful travel plan will need the support and commitment of all members of your organisation. Travel plans can have an impact on conditions of service and, in some instances, staff may interpret the proposed changes as an attempt to reduce their current benefits. It is, therefore, important to involve the Human Resources (HR) department and Trade Union (TU) representatives at the earliest possible stage.

Getting Union and HR staff 'on side'

Enlisting the support of Union officials will be easier if your organisation already has a good track record of consulting and working with the Unions. However, even if this is not the case, development of your travel plan can be taken as an opportunity to develop good Union relations, though it may take longer to make progress.

Developing and implementing a travel plan is a two-way process. Your organisation is trying to make more travel options available to staff but in return expects employees to take up the options, at least some of the time. Much depends on goodwill. Goodwill is often hard to earn and easily lost.

Involving Union and HR staff early in the process will demonstrate that their input is valued. Once involved, they need to be kept informed and an open channel of communication maintained. Clarity and openness about plans will also help to maintain goodwill.

Attitudes towards your plan will depend on your organisation's circumstances and ethos but also on the personal views of individual staff members. The general arguments in favour of developing a travel plan for your organisation will be relevant for all, but some arguments will carry more weight than others, depending on the individual.

Cost savings: Everyone will recognize that cost savings are of benefit to an organisation's bottom line ... However, Union officials will be concerned that savings are not made at the expense of staff benefits.

Investment: If an organisation can demonstrate that the savings will be invested in the other, sustainable aspects of the plan, such as covered bike racks or an incentive scheme for car sharers, it is much more likely to attract support.

Fair treatment for all: If plans include limiting car park use, for example, it is likely to be seen as erosion of a benefit. However, it should be pointed out that by providing free parking to car drivers in the past, the organisation has been subsidising them considerably more than staff travelling by other means.

Energy savings and the environment: Being able to demonstrate savings in energy consumption and the knock-on effect on the local and global environment will usually be seen as positive and deserving support.

Union's position: It is likely that the Union will have a policy on transport issues. Find out more about the stance taken by the Union(s) in your organisation. It is possible that the local representatives may not be fully informed and if you can demonstrate that the principles behind your proposals are in line with Union thinking they may be more supportive ...

Importance of having Union and HR staff 'on side'

Union officials and HR staff will have an essential role to play in all stages of the plan. Their relationship to employees is quite different to that of a line manager and in some cases, a Union official's endorsement may carry more weight than a manager's ...

Getting Union and HR staff 'on side' can play a crucial part in the future acceptance and take-up of the plan. Union and HR staff can be involved in various ways [, including data gathering, ideas development and implementation, for example]:

[...]

- Help to disseminate and conduct the survey (HR/TU).
- Help to ensure the suggestions are fair and realistic (HR/TU).
- Be an initial sounding board for new ideas before they go out to consultation (HR/TU).
- Assist in organising consultation, focus groups, feedback sessions (HR/TU).
- Suggest incentives for take-up (HR/TU).
- Ensure initiatives from 'grass roots' are driven forwards (bottom up ideas are often more likely to be accepted by the workforce) (HR/TU).
- Raise awareness of why the travel plan is being developed (HR/TU).
- Include a travel plan briefing for new employees in the recruitment and interview process (HR).
- Include HR/Union representatives on interview panels when appointing a travel plan co-ordinator (HR/TU).
- Advise on developments elsewhere in the company which could have an impact on the travel plan.

[...]

Influencing change in an organisation's culture

In most cases, transport issues and modes of travel are not central to an organisation's concerns. The ease with which your organisation will adopt a travel plan will be influenced by the kind of culture already in place. Organisations with an open and accessible management style and with effective internal communication structures are likely to be good candidates.

Understanding the way your organisation functions internally, and in relation to those outside, will help to identify the most appropriate ways of introducing change. Likewise, understanding how employees see themselves in relation to the organisation will help determine suitable approaches. Do they feel engaged with the organisation and responsible for their travel choices?

Corporate culture tends to come from the top. It is, therefore, necessary to ensure the most senior executives support the change.

Plan the change

Change is resisted if there is no perceived need for it. A request to complete a Staff Travel Survey can be the first employees hear about impending change. This will often be too late. It is very hard to sell something to someone who does not know they might want it. So, start the debate two or three months before your survey is undertaken. This can be done very overtly with a poster campaign around the premises. This could start with some national statistics and progress to narrower, local facts. These could include statistics on time spent commuting to work ..., costs of commuting by car ... and environmental information ... Include facts specific to your organisation.

Discussion can also be initiated through line management. Use the members of your travel plan Steering/Working Groups to raise the issues in their own departments or make use of team briefings as fora for discussion. Once the ground has been prepared, rolling out the Staff Travel Surveys and site audits will ensure the issue remains in people's minds.

Bringing about change

Knowing what makes your organisation tick is crucial to identifying which measures might be appropriate to your organisation.

Case Study – Langley

Computer Associates in Langley, near Slough, Berkshire, rewards its car sharers who share 26 times in a six-month period with £150 for the first six months, £175 for the following period and £200 for every period thereafter. In a highly paid industry, the company recognised that financial incentives would motivate staff. They were right, there has been a 30% uptake.

[...]

Answering the questions below will help you to devise an appropriate strategy.

- What is going to motivate staff?
- Are senior management supportive? They need to be.
- Are senior management leading by example? They should be.
- Does your organisation already have a good internal communications network? If it does, this will help. If it does not, improvements need to be made. This will benefit the company overall.
- Does your organisation have a consultative style in its decision-making processes? If so, staff will be familiar with the kinds of processes involved. Again, introducing those in the travel plan development process can encourage similar practice in other areas of the organisation.
- Do you fully understand the various communications and decision-making channels in your organisation? You need to.
- Are environmental issues already of interest to your organisation? If they are, travel issues will be readily understood to be part of that.
- Do staff see themselves as responsible for their travel mode or is it seen as 'someone else's problem'? Your travel plan will need to take attitudes and expectations into consideration.

Actions to consider

1 Identify key influencers in your organisation. They may not be the most senior, but they will be respected for their achievements. Enlist their enthusiasm.
2 Identify and enlist support from those that have an 'environmental conscience'.
3 Enlist the support of Union representatives if you have them.
4 Develop a marketing strategy. Use your organisation's specialist staff if you have them.
5 Develop an internal communications strategy, again enlisting specialist help if it is available.
6 Identify if there are other areas of your organisation where changes are necessary or being made. Can you work together?
7 In a large organisation, there may be one or two departments that already have a culture that will make them more amenable to change. Focusing initially on them and being able to demonstrate success there, is likely to make change easier elsewhere.
8 It is important to 'sell' the travel plan to staff at the recruitment stage and to get them on-board and supportive of your plan's objectives as soon as they join the organisation.
9 If you are part of a large, multi-site organisation, with national policies that affect travel and transport issues, you may need to address policy changes with Head Office personnel/management. Head Office should themselves be encouraged to develop a travel plan and lead by example.

[...]

Focus on the areas that will be most relevant to the group you are dealing with. The financial impacts of the plan may be of more interest to the Sales Team than the environmental benefits, for example. Be aware of the pros and cons of the measures that are being planned/introduced and be ready to put the counter-arguments forward.

If you can demonstrate that your organisation is genuinely interested in providing benefits for staff along with benefits for the organisation, they are more likely to respond favourably. Success is very attractive. If staff see an initiative is successful, they will be encouraged to join in.

Energy Efficiency Best Practice Programme, 2002, Sections 2.10 and 2.11

Figure 4.3 A travel plan awareness display for staff

4.5 Travel plan measures

The above section went through the strategic steps and issues involved in setting up and running a travel plan. In this section consideration is given to the specific measures that a travel plan can contain. As was noted at the beginning of Section 4.4, the package of travel plan measures will need to be tailored to each individual site, and the way this happens in practice has already been noted in the hospital case studies in Chapter 3. This section draws upon wider experiences of travel plans to review measures that have been used and how well they work in practice.

Encouraging travel to work by train or bus

Transfer to public transport is, as was noted in Chapter 1, widely espoused as an environmental measure. The extent to which this is viable for individuals varies greatly, as does its appropriateness for employees and visitors to specific sites. City centre sites are likely to be better served by public transport than other areas, but the example in Chapter 3 of Derriford Hospital in Plymouth, on a suburban site, shows that it is possible to significantly improve public transport access elsewhere.

There are several ways in which employers can encourage their staff to get to work by public transport. For instance, they can:

- provide public transport information in the workplace
- negotiate public transport discount from bus or rail operators to either enhance services or reduce fares
- subsidize public transport (to enhance either services or fares)
- provide 'works buses' to supplement existing public transport, or
- promote rail for business travel.

There are several examples of successful train and/or bus measure in travel plans. One comes from Buckinghamshire County Council, who negotiated significant discounts for staff to use local public transport. As a result, staff paid half fare on Arriva buses and got a third off rail fares on services operated by Chiltern Railways. Both operators attracted enough new custom to profit from the arrangements, and public transport use among County Council employees increased from 8% to 14% (Department for Transport, 2002b).

Stepping Hill Hospital in Stockport negotiated a discount with local bus and train operators of approximately 5% for staff displaying employee travel cards. This may be further subsidized by the hospital to give a 20–30% discount, using revenue raised from car park charges, as is done at Derriford Hospital (NHS Estates, 2001). In addition, as in Nottingham City Hospital, also considered in Chapter 3, buses ran onto the site to drop off and collect passengers.

Egg, the eBank based in Derby, introduced several measures. A public service shuttle bus, subsidized by Egg and used by 14% of staff, ran every 12 minutes between its site and Derby bus station. While the service was initially free to staff, a nominal 10p fare was introduced later. Also introduced was a free contract bus between Egg and a nearby park-and-ride site. In addition, liaison with the local authority led to the installation of two new bus stops and shelters close to site entrances (Department for Transport, 2002b).

The mobile phone company Orange funds a fleet of six single-deck buses to operate on two routes between Aztec West and Almondsbury business parks in the north of Bristol and its new Temple Point office in the city centre.

Figure 4.4 Like Orange, the pharmaceutical group Pfizer operates works buses that connect its site to the local railway station

BAA, the operator of London's Stansted Airport, had an impact on the local buses that extended a considerable distance from its site. This was because it used its travel plan to address a problem of staff recruitment. With a shortage of staff locally, the company sought to recruit people living further away along public transport corridors, particularly the rail corridor into London. This resulted in money being spent to improve the quality of the 123 bus route that links Ilford and Wood Green in north and east London to Tottenham Hale station (the only stop on the Stansted Express service from London's Liverpool Street station). Thus people travelling on the 123 bus in London can thank BAA Stansted for an enhanced service.

An important development is that the growth in travel plans has begun to produce initiatives from bus operators. For example, bus company First Hampshire has targeted a thousand companies in its area with details of its travel plan scheme. The *Take One to Cure Congestion* leaflet (designed to resemble a packet of aspirins) is aimed at raising awareness of the company's existing bus network, as well as explaining how it can plan and operate bespoke staff shuttle bus services for employers in the area. Similar schemes have already been established with Portsmouth City Council and with the main hospitals in both Portsmouth and Southampton (*Transit*, 2002).

Reducing the need to travel

One obvious way of reducing parking and traffic problems is to reduce the need for making journeys in the first place. Reducing transport dependency was identified in Section 1.8 of Chapter 1 as a crucial component to cut energy and emissions from transport to a sustainable level. One travel plan measure that takes this approach is the introduction of flexible working arrangements that permit employees to travel a little earlier or later than normal to avoid the busiest time on the road, thus saving time and leading to some reductions in fuel consumption and emissions due to better driving conditions. The impact upon energy use and emissions of such practices is marginal, but other flexible working practices can have a substantial impact.

One example is 'compressed working', which may involve people working, say, four-day weeks or nine-day fortnights, but with longer days. In California, the city of Irvine introduced a compressed working week in 1991. During the first nine months, not only did this cut the amount of commuting and pressure on parking spaces, but there was also a 16% reduction in sick leave and a 17% reduction in staff overtime worked compared with the same period the previous year (Department of the Environment, Transport and the Regions, 1999). This illustrates the indirect (and significant) benefits to an organization that travel planning can achieve. Another example is BP's office at Sunbury-on-Thames, where staff are encouraged to work slightly longer days in return for a day off each fortnight. Pfizer, in Kent, also offers staff a compressed, nine-day fortnight. There are also plans to set up satellite offices in areas where staff live, overcoming potential problems of isolation for those working at home (Department for Transport, 2002a).

Chapter 1 also looked at teleworking, whereby people work from home using communication networks. The widespread use of such flexible working practices can also permit 'hot desking', where people share workstations rather than have one each, which may be under-utilized. Alternatively, occasional work spaces can be provided in company sub-offices. If flexible working practices reduce the amount of office space required, very substantial savings indeed can result, especially in high-cost city centre offices.

Buckinghamshire County Council has promoted home working and 'hot desking', where possible, to reduce travel, and at the Government Office for the East Midlands in Nottingham, flexitime is encouraged and laptop computers are available for staff to use at home or on public transport. Boots is another company that has encouraged home and flexible working practices to be more widely adopted, and AstraZeneca staff can apply to have a web camera on their PC or laptop (Department for Transport, 2002a).

Such approaches are usually popular among staff, while the costs to the employer can be minimal and can even result in large savings. However, there are implications for administration. Management is also often concerned about staff supervision, although this is viewed as a rather old-fashioned attitude. According to the *National Travel Survey 1998–99*, in Britain about one million people (or 3.7% of the workforce) usually worked from home, and a further two million used their home as a work base but also worked elsewhere (Department for Transport, Local Government and the Regions, 2001a).

One reason that people often drive to work is because they need to visit a bank or go shopping during their lunch time, a trip that would not be possible without a car. Providing on-site services, such as a shop, chemist, newsagent or cash dispenser, can therefore help, particularly in larger, more isolated locations. Even if people still drive to work, such measures cut down on the amount of driving needed. For example, at The Open University site in Milton Keynes, a van from a local Waitrose supermarket used to deliver pre-ordered goods at the end of the working day to a car park where staff could load it straight into their cars. Some companies, particularly those in out-of-town locations, also operate free or subsidized 'works buses' for lunchtime shopping trips. A combination of these measures is implemented at the DVLA offices at Swansea, which have an on-site pharmacy and a dry-cleaning collection service, and also operate a lunch-time shopping shuttle bus service into the town once a week (Department of the Environment, Transport and the Regions, 1999).

Reducing business travel can be integrated into a travel plan and can result in substantial cost savings to an organization. New technologies play an important part in enabling a change in travel behaviour. Information technologies such as the Internet, teleconferencing or phone conferencing can remove the need for trips altogether and assist home workers. Several companies are beginning to expand their use of videoconferencing to cut business travel. These include Vodafone in Newbury, Egg in Derby, the Government Office for the East Midlands in Nottingham, and BP at its Sunbury-on-Thames office. AstraZeneca has set up six to eight videoconferencing studios at its site in Macclesfield (Department for Transport, 2002a).

Car park charges and cash-out

The provision of free or cheap car parking is, in practice, a subsidy provided by an organization only to those who drive. Indeed, the availability of a free car parking space is one of the main influences an employer has on people's travel behaviour. Surface car park construction costs £400–£800 per space plus annual maintenance of £100–£500, while the cost of building each multi-storey or underground space is in the region of £6000 (Energy Efficiency Best Practice Programme, 2001). Consequently, introducing parking controls, restrictions and/or charges or paying staff to give up their parking space can be very cost-effective. But the very effectiveness of charging staff for parking also often makes such actions extremely unpopular and difficult to introduce.

An alternative, which is obviously more acceptable, is to pay drivers not to use their cars for certain trips – effectively bribing motorists to use an alternative mode. One application of this idea, the 'parking cash-out', is in use in the UK.

As noted in Chapter 3, Derriford General Hospital in Plymouth has a parking cash-out scheme applying to staff who regularly commute by car three or four days a week. Applicants are monitored over a four-week period to see if they qualify. If they do, they are then given a one-off payment of £250, plus an extra amount to cover VAT. In return, staff members forgo their right to park by handing over their parking permits and having their ID codes erased from the parking monitoring system. The scheme was introduced in mid-2000, but by 2003 only seven people of the 3500 (0.002%) who qualified for a parking permit had taken up the benefit (although 25–30 people had applied). In 1997 airport operator BAA offered employees £200 each to forgo their parking spaces at Heathrow. This was a little more successful than at Derriford Hospital, with 33 (around 1%) taking up the one-time offer.

Rather than giving just a one-off cash-out, a scheme started in 1995 at Southampton General Hospital gives car park permit holders an initial payment of £150 and subsequent annual payments of £96. Take-up is larger than at BAA or Derriford. As of autumn 2001, 551 out of 5911 permit holders (9%) had taken up the scheme. A monthly, rather than annual, system is in operation at the Vodafone offices in Newbury, Berkshire. Introduced in 2000, the scheme allows any employee to opt out of having a parking space and receive an extra £85 in their monthly pay packet. This substantial incentive has resulted in 1500 (a third) of the 4500 staff based in the town taking up the scheme.

From the above examples of parking cash-out, it appears that there is a pattern of take-up related not only to the amount of money offered, but also to the degree of flexibility involved. It is one matter to say that you will not be able to drive to work for a month and then review the situation, but quite another to say you will never drive to work again! Furthermore, if a scheme is inflexible employees who might feel happy not to drive one, two or three days a week cannot benefit because they need to use the car on the other days.

To address this problem, in 2001 the pharmaceutical giant Pfizer introduced a flexible parking cash-out scheme that rewards non-car commuters on a daily basis at its research and production facilities at Sandwich in Kent and at Walton Oaks near Reigate in Surrey. This works by using a staff-personalized security pass involving 'proximity card' technology. An employee's card is credited with enough points to 'pay' for one month's parking. The card opens the parking barriers and records how many points are used. At the end of each month staff cash in any points they have not used for parking, payments being made through the payroll. Staff at the Sandwich site receive £2 per day for leaving their car at home, while at Walton Oaks the incentive is £5 per day – a reflection of the far tighter parking standards set by the local planning authority. Overall, it is estimated that the value of cash-outs given to staff will cost Pfizer around £0.5 million per year. The impact upon travel choice is, however, substantial. In 2003 around a third of staff travelled to work other than by car to locations that would normally be very car-dependent.

Figure 4.5 Using the Pfizer 'proximity card' pass to pay for parking at the company's Sandwich site. Staff can collect £2 per day if they leave their cars at home

The Pfizer example illustrates how much easier and more effective it is to persuade people to switch from using the car for one or two days a week than for four or five days a week (or for ever).

The travel plan measures that allow people to regain their right to park if they have a baby or move house, or ideally to park when the weather is bad and cycle when it is dry and sunny, are likely to be more appealing. This shows the care that is needed in developing a travel plan measure. Being stingy will yield little benefit for the costs incurred. Parking cash-out can be an expensive travel plan measure, but may not be as expensive as providing parking. At an annual cost for surface spaces of at least £500 and for multi-storey spaces of £6000, an annual cash-out payment of £500 can be a cost-effective alternative to car park construction. But it will also require close co-operation from a local council to control parking in streets near a site, to prevent staff from abusing the system (by taking the cash reward and then parking nearby). Nevertheless, providing financial incentives to drivers, and existing commuters who already do not drive to work, certainly generates far less staff opposition than introducing charges or just restricting parking spaces.

Walking and cycling to work

As nearly 60% of all car journeys to and from work are less than five miles long and a quarter are less than two miles, promoting walking and cycling can play a significant role in reducing car trips. Travel plan measures that can help encourage walking and cycling include:

- promoting the health benefits of walking and cycling to work in staff newsletters
- identifying safe routes to the workplace and publicizing them in maps or guides
- providing lockers, showers, changing rooms and secure cycle parking
- negotiating with the local authority to build safer off-site walking and cycling routes
- granting interest-free loans and/or discounts for the purchase of bicycles.

The sustainable transport plan prepared by the Stepping Hill NHS Trust in Stockport included a number of measures to promote walking and cycling. Specifically, it introduced 'green' route maps and newsletters, built showers and changing facilities, and improved cycle parking security. It also negotiated discounts with local bicycle shops, offered a 'bike doctor service' on site for bicycles that needed repairing or maintaining, and mounted awareness-raising events. Finally, from the summer of 1996, the Trust purchased 85 bicycles for staff to lease, with a commitment to buy a further 25 each year. The scheme was funded from car parking revenue (Collins, undated).

In another of its travel plan cash incentives, the Langley-based business software company Computer Associates (see the short case study in Box 4.2 above) gives staff a £150 cash incentive if they walk or cycle to work 25 days in six months. Members of the walking scheme receive a free pedometer and a walking route map. The company also provides a fleet of 25 company bicycles to ride to and from work with free accessories, a state-of-the-art secure cycle shelter, and lockers, drying facilities and showers. As a result, some 12% of staff signed up to the cycle-to-work scheme (Department for Transport, 2002b).

Car sharing/lift sharing

In the context of travel plans, the term 'car sharing' usually refers to offering lifts to work, school or college. How this is done ranges from informal lift-sharing arrangements among friends within one business or street, to formal arrangements using computer databases. In general, employers can help staff by establishing a car-sharing database, giving parking priority to car pool vehicles, charging car pool vehicles less to park, and setting up a guaranteed ride home scheme to cover emergency events for lift sharers. Sharing a car where people are attending the same meeting can also be promoted as a way of reducing single-occupancy car trips for business journeys; an example is the provision of a car passenger mileage rate for business trips to reward motorists who transport colleagues to business meetings.

Over 30% of employees working at Marks & Spencer Financial Services in Chester now car share one or more days a week. Sharers are matched using a computer database, are offered the most convenient parking spaces at the front of the building, and are guaranteed a lift home if arrangements fall through. A range of financial incentives also encourages staff to car share. Those joining the scheme receive an M&S voucher worth £20, while those completing six months can choose from a range of car-related perks – the cost of road tax (to the value of the lowest UK charge band) or the same amount of money spent on car servicing or petrol vouchers. Those completing 18 months receive M&S vouchers worth £50 (Department for Transport, 2002a).

The brief case study of Computer Associates in Box 4.2 above also illustrated how cash incentives can be used to promote car sharing. As in the case for parking cash-out, it is important to provide users with flexibility.

(a)

(b)

Figure 4.6 (a) Reserved car share parking spaces at Boots Beeston site, Nottinghamshire. (b) the car sharer's special parking permit

Potential car sharers at Buckinghamshire County Council can find matches through a centrally co-ordinated scheme. Four prize draws a year encourage participation, while funds are also set aside for a guaranteed taxi ride home should lift arrangements fall through. Car sharers are exempt from parking charges and can use a 'green bay' space in a nearby car park. Similarly, the Derby-based internet bank, Egg, exempts car sharers from a 75p per day parking charge; in 2003 the proportion of staff sharing cars was about 25%. Publicity for the Buckinghamshire car share scheme emphasizes financial savings; for example, one group of sharers were able to use the money they saved to go on holiday (Department for Transport, 2002b).

Car sharing is also promoted at Agilent Technologies just outside Edinburgh, where cars carrying three or more people are able to use dedicated 'green bay' parking spaces located in prime areas. Car sharers initially found matches on a notice board but the service is now available on the company intranet. Usage has nearly doubled in five years and regular users from further afield (for example Glasgow) claimed that they saved around £100 a month (Department for Transport, 2002a).

4.6 Freight mobility management

Travel plans in the UK tend to be associated with measures to cut the transport impacts of individuals journeying to and from a site, be they employees, school children, customers, students or football fans heading to a stadium (yes, several UK football clubs have 'fan travel plans'). But the transport of goods and deliveries in general can also account for a substantial part of the transport impacts of an institution. Mobility management can be applied to freight as well as to people. An example of this is provided by the following case study of Heathrow Airport's Retail Consolidation Centre, which is taken from a UK Government Energy Efficiency Best Practice Programme case study.

BOX 4.3 Heathrow Airport Retail Consolidation Centre

'When we started this project our vision was to create a 21st Century logistics operation which would overcome the airport's physical constraints and allow Heathrow Airport Limited to grow its retail business and fulfil its environmental objectives.

To achieve our goal we needed a radical change to our delivery process, but recognised the need to involve our customers from the start if we were to succeed. We have developed an excellent working relationship with our logistics partner, Exel. They share our enthusiasm in not only making the current operation a success but in developing the product to extend to other parts of the airport's business.'

Brian Gibb, Security Development Programme Manager, Heathrow Airport Limited

Introduction

[...]

Retail development within Heathrow airport has increased dramatically over the past ten years, but the infrastructure has seen little change to accommodate this growth. Congestion, both on airport roads and at loading bays, was a significant problem with 439 supplier movements to 240 retail outlets being made each day. Clearly with the development of Terminal 5, this congestion would only increase.

BAA's challenge was therefore to find a new distribution strategy that improved efficiency and reduced congestion and pollution. Their solution was to establish an off-airport consolidation centre, managed by the logistics provider Exel on their behalf.

[...]

The need for change

The delivery operation across Terminals 1–4 had evolved over several years and was no longer adequate to support the growing retail business at Heathrow. Particular problems included:

- An overloaded central terminal area and single tunnel access.
- Poor infrastructure.
- A lengthy and unpredictable delivery service.

These issues, together with growing environmental pressures and the potential requirements of Terminal 5, demanded a radical re-think of the previous operation.

The planned new Terminal 5 is expected to add another 450 000 sq ft of retailing space within the airport, which is approximately equivalent to 250 new retail units. If Terminal 5 were to operate on a similar basis to that used previously at Terminals 1–4, it would require 64 new delivery bays plus a substantial parking area. A lack of space in the proposed site for Terminal 5 prevents this from being a viable option.

In anticipation of these plans for Terminal 5, Senn Delaney, part of Arthur Andersen & Co., performed a study of truck movements at Heathrow in 1996.

They produced various proposals as to how the number of vehicle movements supplying retail units at Terminals 1–4 could be reduced. The study evaluated various retail delivery options and reviewed both the planned infrastructure and new methods of approaching the delivery problem. Some of these solutions were taken from other retail practices world-wide. It concluded that the best combination of traffic volume, physical infrastructure requirements and delivery/handling costs would be met by the creation of a consolidation centre. The study recommended that this consolidation centre be located away from the airport and that all retail merchandise and catering foodstuffs be delivered to it ...

After accepting the findings from the consultants' report, BAA sought to develop an alliance with a key logistics operator in order to manage the supply of goods to the retail outlets in the airport. This partnership had four main aims:

1. To improve methods of delivery to retail units
2. To reduce vehicle movements through consolidation of products
3. To improve handling at delivery point both on and off the airport
4. To improve management of packaging waste

Four month trial

Between March and June 2000, Exel performed a trial of this off-airport consolidation method. Eight retailers volunteered to participate (including Bally, Tie Rack, City Centre Restaurants and Sunglass Hut / Watch Station) corresponding to some 40 retail outlets across the four terminals. The trial produced a 66% reduction in deliveries to the airport and good feedback from the retailers. The consolidation centre allowed more flexibility of delivery times, and staff did not have to leave their shops to collect packages from a central point.

Other benefits included:

- A quicker turnaround of delivery vehicles at the consolidation centre
- Less pressure on hauliers regarding drivers' hours regulations
- Better vehicle utilisation
- Easier start-up for new retailers within the airport
- Predictability/reliability of deliveries within agreed windows to both the consolidation centre and retail outlets
- Less in-store stock required due to quick, reliable replenishment

Following the success of the trial logistics companies were invited to submit proposals for the large-scale delivery of the scheme. The five-year contract worth £2 million per year was awarded to Exel starting in May 2001.

Operation

Exel currently operate a 25 000 square ft warehouse at Hatton Cross with five vehicles (4 × 17 tonne rigid box-vans with tail-lifts and 1 × 3.5 tonne with tail-lift) and some 20 operational and clerical staff. Freight arrives at the centre in a variety of packaging from the suppliers. It is subject to security checks, caged and sealed ready for despatch. The seal is part of a DfT approved security arrangement that includes a Rapiscan X-ray machine, operated by trained Exel staff ...

A chilled storage facility (3500 sq ft) became operational in November 2001 allowing the consolidation centre to meet the specific requirements of catering outlets by delivering chilled and frozen products. Previously, these products were stored in insulated containers (Thermotainers) that could be transported into the airport using non-specialist vehicles.

[...]

When the Exel vehicle unloads at the delivery bay within the airport, an Exel employee brings the goods to the store (or stockroom) in a roll-cage together with delivery notes from both the supplier and Exel. Any stock that needs to be transferred between branches within the airport can be labelled with the new delivery address and will be delivered on the next visit to the named branch.

A driver and usually two terminal staff man [*sic*] each delivery vehicle, which remains at a terminal until deliveries to all the retail outlets are completed. Security checking procedures have been embedded into the system at the consolidation centre, reducing the number of checks that need to be made for vehicles destined airside. All the delivery and terminal staff are screened for security.

Deliveries can be scheduled to suit the retailers' preference. The consolidation centre is now open 24 hours a day, 7 days a week for deliveries and also provides onward transportation and delivery to the airport at any time specified by the retailer. Exel also collect and remove recyclable packaging waste, such as cardboard and plastic. The delivery staff are trained to ensure that they are aware of the environmental role that they can play.

Some retailers also use Exel for the provision of additional services, such as:

- Overnight stocking
- Carrying a buffer stock that can be picked and delivered on demand
- Providing stock levels/records
- Receiving in bulk and breaking down for multiple deliveries
- Pre-retailing, for example, point of sale displays
- Seasonal stock-holding

Retailers are charged per roll-cage delivered at a rate based on a cost-neutral position for the retailers, in terms of the savings brought about by consolidation. This process was adopted to maximise participation of retailers in the scheme during the first year of operation. Additional services are charged for separately. Exel have an 'open-book' contract with BAA and charge a management fee for their services and expertise.

Benefits arising from the consolidation centre

In summary, Heathrow's new consolidation centre has achieved:

- A reduction in the number of vehicles travelling to the terminals, thereby reducing congestion both within the airport and on the approach roads. These improved traffic flows within the airport benefit all airport users both in terms of reduced congestion and improved air quality.
- A reduction in the number of vehicles passing through control points and driving airside, thus reducing the number of less experienced drivers on these airport roads. Flight operations (e.g. aircraft refuelling, baggage handling, flight catering) have been greatly enhanced by this reduction in congestion.
- Faster deliveries being made to the consolidation centre by suppliers since there is less congestion on approach roads and fewer delays in off-loading goods and loading returns.
- More frequent and scheduled deliveries to the terminal buildings, enabling retailers to know more accurately when goods will arrive, within agreed delivery periods. This helps a retailer to receive merchandise in a shorter time, something that is greatly appreciated.
- Potential savings in both supply chain and staffing costs for retailers using the consolidation centre.

Environmental analysis

In the last week of January 2002 there were 115 inbound deliveries to the consolidation centre ...

These 115 vehicles per week now delivering into the consolidation centre would previously have all delivered into Heathrow airport itself. Excluding the journeys made by Exel from the consolidation centre to the airport, 35 fewer trips were made into the airport that week. Since each journey is 10 miles, this comes to a saving of 350 miles per week. Assuming that the suppliers' vehicles have a fuel consumption of 10 mpg, then 35 gallons (159 litres) of diesel fuel are saved, worth £100 a week or £5125 per year to that single supplier. This corresponds to a weekly saving of 426 kg of Carbon Dioxide ...

This reduction in journeys also saves 1.35 kg of Carbon Monoxide, 1.06 kg of NMVOC (Non Methane Volatile Organic Compounds), 3.79 kg of Nitrogen Oxide and 0.28 kg of Particulates a week.

Since these savings were calculated at current business levels with the consolidation centre serving only 40% of potential outlets, they are likely to be increased further in the future. Emissions are also expected to reduce when a new alternative-fuelled fleet is introduced. The intention is to use compressed natural gas fuelled vehicles, as they provide the best current improvement to local air quality.

Future developments

Exel are currently introducing a computerised hand-held tracking system, based on Palm Pilot technology, which uses software specifically written by Exel for the BAA operation. Every carton will be given a bar-coded label, as will every cage so that goods will be scanned in and out of the consolidation centre. Load

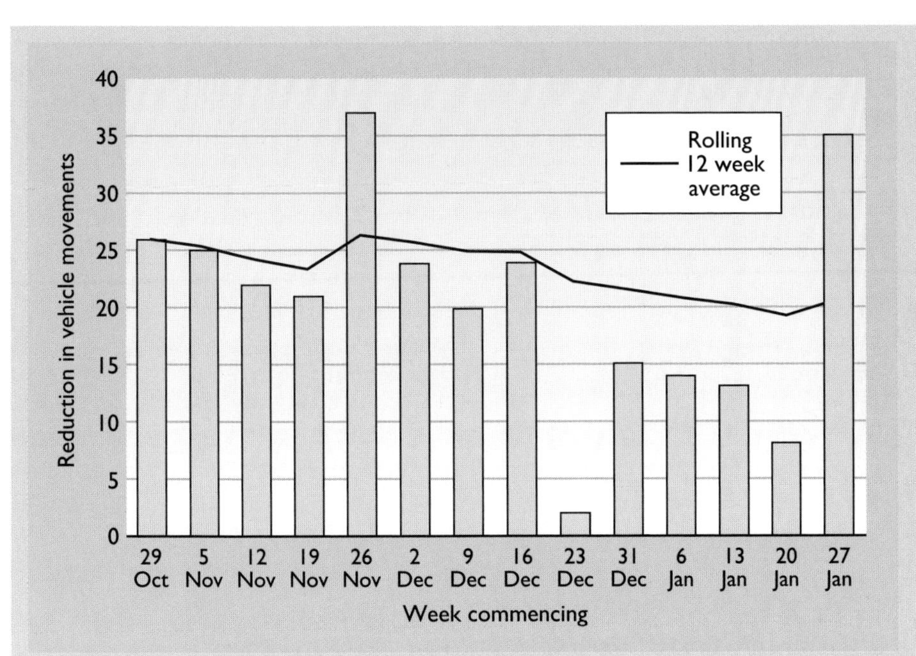

Figure 4.7 Reduction in vehicle movements Oct 2001 – Jan 2002

planning will be performed electronically, producing a full manifest by cage. Each cage will be sealed and tracked as it moves around the airport and will also be scanned at retail outlets. This development will increase the efficiency of this delivery operation still further.

Currently the Duty and Tax Free shops, operated by World Duty Free plc (a subsidiary of BAA plc), are not included in the consolidation centre operation. They are serviced by a separate distribution facility in close proximity to the airport that could be incorporated into the Hatton Cross facility [in] the future ...

All new concessionaire agreements and renegotiated existing agreements at Heathrow will require retailers to use the Exel facility, so that by the end of 2004 all retailers will be incorporated within the system as existing franchises are renewed.

There will also be a study of the possible efficiencies of supply chains both within and between other BAA owned airports. The consolidation centre currently services Gatwick Airport three times a week, and London City and Stansted Airports. BAA is keen to examine how other airport operations in UK could benefit from the use of a consolidation centre in their distribution operations.

Conclusions

The BAA consolidation centre, run by its chosen logistics partner Exel, was awarded the 2001 Institute of Logistics and Transport Environmental Award for its contribution to the environment. It has already produced significant environmental benefits and is regarded as a blueprint for future retailing operations, both at airports and in other locations.

At the start of the project, BAA was under increasing pressure to provide a solution to the airport's physical constraints. Some 400 vehicles were delivering goods on an unscheduled daily basis to retailers, causing massive congestion within the airport and on the approach roads. Off-site consolidation was the solution chosen to solve this problem, with the operation commencing in May 2001. The project has been very successful so far, and it is planned to include all retailers operating within Heathrow by 2004.

The partnership between BAA and Exel contributes to the environmental strategy for Heathrow by identifying base data, measuring, monitoring and setting targets to demonstrate the improvement in air quality and packaging waste management – as well as service levels and ease of access.

Heathrow airport has seen a significant reduction in the number of vehicle movements as a result of this scheme. On time delivery performance to the retail outlets is currently 95%. BAA have been able to set targets at full implementation of a 75% reduction in the number of vehicles delivering to the airport and a 90% use of vehicle load capacity.

Retailers operating within the airport receive more effective, on-time deliveries on high security shared-user vehicles. The project has been so successful that any new retailer uses the consolidation centre as a condition of contract. With such positive commercial and environmental benefits, this type of solution may be adopted not only by other airports but also by retailing operations with similar congestion problems such as those based in city centre locations.

Future Energy Solutions, 2002

Heathrow Airport also has a travel plan for its staff and those of other on-site employers. This case study shows the importance of extending the concept of mobility management to cover the delivery of goods when, as here, they form a major part of movements to and from an organization's site.

For Heathrow it is notable that the key stimulus to freight mobility management being considered was on-site congestion. This has parallels with travel plans for people, where parking and congestion problems are often a key factor behind the introduction of mobility management measures. It is also notable that the consolidation centre produced cost savings to the airport and the delivery companies and a better service to the retail outlets. Yet had it not been for the congestion problem, presumably the existing, inefficient delivery system would have continued. As with travel plans for people, the normal institutional response fails to appreciate the gains in efficiency that are possible. Having realized the commercial and operational benefits of Heathrow's consolidation centre, BAA is now looking to apply the concept at other airports where congestion in deliveries is not a particular problem.

The consolidation of deliveries is fairly widespread as a freight mobility management measure. Another example is at Tulln Hospital in Austria, where the initiative arose not because of a site congestion problem, but through a general management review to improve the environmental performance of the hospital that was linked to obtaining environmental management certification (ISO 14000). As part of this, the hospital consolidated minor deliveries from a range of suppliers into one collection and delivery vehicle. It also co-ordinated pharmacy deliveries through a single supplier, and also consolidated waste disposal, including recycling collections. The result of this mobility management strategy was a reduction of about 23% in transport associated with delivery of materials.

The Heathrow case study also shows how technical improvements to reduce environmental impacts can be integrated with a mobility management strategy. Although the consolidated deliveries were initially undertaken using conventionally powered vehicles, the intention is to convert to compressed natural gas (CNG) vehicles. As was noted in Section 2.7 of Chapter 2, CNG has lower emissions of pollutants, particularly NO_x and particulates (although, for heavy-duty vehicles, there is little difference in CO_2 emissions compared with diesel). CNG has limited use in the UK, because it requires special gas compressors for refuelling. By creating the consolidation centre, served by a self-contained fleet, BAA Heathrow created the conditions that allowed the use of CNG. This small but intensively used fleet of vehicles can justify investing in a gas-compressor refuelling facility. Thus a mobility management strategy has made possible a fuel-switch strategy as well.

4.7 'Greening' transport – a difficult challenge

Chapter 1 started by identifying the major challenges to reducing the energy use and environmental impact of transport activities. The difficulty of achieving sustainability for transport was demonstrated by use of a simple index model to explore current trends. The model indicated that a combination of technical advances and behavioural changes is necessary if the personal transport sector is to achieve a sustainable level of CO_2 emissions, in order to meet climate change targets in the medium to long term. The chapter suggested a possible 20-year strategy involving:

- a 30% increase in car and bus journeys (rather than 50% rise by car in the 'business as usual' scenario)
- halving trip lengthening except for rail (assuming this will pick up some long car trips)
- a 50% improvement in car fuel economy and 40% for bus and rail
- a 20% cut in the carbon intensity of the fuel used for road vehicles and a 30% cut for rail
- modal shifts as suggested by the Royal Commission on Environmental Pollution (1994), cutting car use from 88% of motorized trips to 65%, with bus use rising to 25% and train use to 10%.

This is a rather radical strategy, and when one looks at the current political debate over transport issues in both the UK and other developed (and developing) nations, one may well question whether such progress is likely to be achieved. Chapters 2 and 3 examined the elements of this sustainable transport strategy and explored what progress appears to be likely.

Road transport technologies

Chapter 2 looked at technological improvements to produce better fuel economy and cut carbon intensity. The key question is whether such technological improvements could yield the 50% improvement in car fuel economy and 20% cut in fuel carbon intensity that would keep us on track towards a sustainable transport system.

In terms of emissions from road transport, there have been substantial reductions in pollutants such as CO, NO_x, particulate matter and volatile organic compounds. Already technological developments have produced an improvement to air quality in towns and cities where traffic fumes have been a serious problem.

However, in terms of fuel economy and CO_2 emissions, progress is less encouraging. It was noted that trends towards more powerful vehicles and additional power-consuming features in cars have counterbalanced improvements in engine designs. The test fuel economy of new cars shows very little improvement, despite an EU target to reduce CO_2 emissions from new cars from the current 180 g per km to an average of 140 g per km by 2008 and 120 g per km by 2012. The 33% improvement called for by the EU is technically possible, but it seems that, at present, most people do not want to buy fuel-efficient cars. Hybrid cars (Section 2.5 of Chapter 2) might be a way to overcome this difficulty, as they can yield something like a

30% fuel economy improvement while still delivering more power for the type of car people want. However, cost is an issue, with the price of hybrids significantly higher than ordinary petrol-engined cars. The 'moderate technical progress' scenario in the report *Fuelling Road Transport* (Eyre *et al.*, 2002) envisaged hybrids taking a 10% market share by 2012 if they were adequately promoted.

Some alternative fuels, LPG in particular, are beginning to establish a niche market in the UK. These offer a good reduction in regulated emissions of CO, NO_x and particulates, and about a 10–15% reduction in CO_2 emissions compared with petrol-engined cars (5% compared with diesel). Some biofuels (Section 2.8 of Chapter 2) can offer substantial reductions in CO_2 emissions, but there is a wide variation. Biodiesel (rape methyl ester, RME) can cut greenhouse gases by up to 50%, and can be easily distributed, but there is limited land for production.

Overall it is likely that, in the next 20 years, niche markets will open up for a variety of alternative fuels. Infrastructure constraints will probably restrict CNG to fleet uses, whereas LPG and biodiesel could see more general, but not widespread, use. The report *Fuelling Road Transport* (Eyre *et al.*, 2002) includes a 'high biofuels' option, but even under this it is envisaged that by 2020 biodiesel would account for only 5% of all diesel fuel and ethanol for only 5% of petrol consumption. The report indicates a limit to the use of LPG, suggesting no more than 3% of new car and van sales by 2030. The 'moderate technical progress' scenario envisaged 10% of buses using CNG by 2010 and a steady increase of CNG heavy goods vehicles (to 30% in 2050).

All this suggests a more diverse mix of fuels for road transport vehicles than we have at the moment. In 20 years alternative fuels may together amount to, at best, 20% of all transport energy consumed. So, even if these technologies can deliver, say, a 30% cut in CO_2 emissions, because of their limited market penetration this will produce an overall cut of only 6%.

The situation for battery electric vehicles (BEVs), covered in Section 2.9 of Chapter 2, is similar. Like other alternative fuels, they offer substantial cuts in regulated emissions, but only a modest reduction in CO_2 emissions (about 25%, although this would improve over time if more renewable energy were used for electricity production). Given the high cost and poor technical performance of BEVs, it seems they are unlikely to have a long-term future, particularly as fuel cell vehicles, which were discussed in Section 2.10 of Chapter 2, are emerging as the main challenger to petrol and diesel internal combustion engine vehicles. Depending on the primary fuel used to manufacture the hydrogen, fuel cell vehicles can reduce CO_2 emissions by up to 60% compared with a similar petrol-engined car. This would be higher were more renewable energy used for hydrogen production, but (as discussed in Chapter 1) this seems unlikely for road transport use for some decades.

The growth of fuel cell vehicles is the great unknown in looking at road transport futures. Like BEVs they will be expensive to buy and their fuel infrastructure could in some cases be problematic. In *Fuelling Road Transport* the 'moderate technical progress' scenario has 3% of new cars using hydrogen fuel cells by 2020, rising to 30% by 2050. Its 'rapid progress' scenario has 5% using fuel cells by 2010, 20% by 2020 and 100% by 2050.

The best possible cut in CO_2 from fuel cells is about 60%. So, taking the 2020 'rapid progress' figure, with fuel cell vehicles representing 20% of cars, this would result in an overall cut in CO_2 emissions from all cars by 12%.

Adding the above effects together suggests that all the alternative fuel technologies might manage about a 20% cut in carbon intensity per vehicle-kilometre over a 20-year period. The figure in the index model for a 20% cut in CO_2 looks possible, but not easy. However, improvements in fuel economy are another matter. In introducing low carbon fuels, the main aim has been to cut local air pollution, although some incidental fuel economy improvements have arisen. Better fuel economy is technically possible, but requires an acceptance of a change in the size and performance of cars. At the moment, few people are willing to buy fuel-efficient cars. Policymakers have ducked this issue, and favour the less politically contentious option of fuel switching.

(a)

(b)

Figure 4.8 Fuel switching ((a) an electric Peugeot 106) seems an easier option than promoting car designs with good fuel economy ((b) the E-Auto design), but both technologies are needed for transport to become sustainable

Mobility management

Approaches involving behavioural change and mobility management formed the central focus of Chapter 3 and this chapter. Both chapters looked at travel plans introduced by institutions such as employers to manage the amount of travel and the modes of transport of their staff, customers and other users. This chapter also considered freight mobility management.

Travel plans are only one part of a mobility management strategy. As they are implemented by institutions they largely affect journeys to work and business trips, or trips by users to a site (hospital patients and visitors being a good example). These account for, at most, about a third of all travel. Other mobility management measures are needed to address other trip types and travel in general (for example, financial and tax measures such as congestion charging mentioned in Section 1.6 of Chapter 1). There also need to be expenditure measures by central and local government – such as improving public transport or building safe cycle routes – to increase the viability of travelling in a more sustainable way.

Travel plans, together with other mobility management measures, could have an effect in achieving some of the suggestions made in Chapter 1 to reduce CO_2 emissions to a sustainable level. These were:

- cutting from 50% to 30% the increase in the number of car and bus journeys
- halving trip lengthening
- achieving modal shifts, cutting car use from 88% of motorized trips to 65%, with bus use rising to 25% and train use to 10%.

The hospital case studies in Chapter 3 and the wider consideration of travel plan measures in this chapter provide evidence of the impact such measures can have on these three factors. If travel plans are well managed, they can be an effective tool for reducing car use. A study that pulled together evidence on the impact of travel plans (Department for Transport, Local Government and the Regions, 2001b) found that travel plan effectiveness varied depending on the measures that were implemented. Specifically, it found that:

- a plan containing only marketing and promotion was unlikely to achieve any modal shift;
- a plan with car-sharing measures may achieve 3–5% reduction in drive-alone car commuting;
- a plan with car sharing, cycling and large discounts (more than 30%) on public transport plus works buses will achieve a 10% reduction in drive-alone car commuting;
- the combination of the above measures, together with disincentives to drive, can achieve a 15–30% reduction in drive-alone commuting.

Most travel plans, however, do not progress beyond the less-effective levels. They have not been widely adopted, and they are introduced with some reluctance. Rye (2002) estimated that travel plans removed just over 150 000 car trips per day from British roads each working day, or 1.14 billion kilometres per year. This is not a lot: it equates to around three-quarters of one percent of the total vehicle journeys to work. Rye goes on to suggest that this fairly low take-up thus far is due to five major factors, namely:

- companies' self-interest and internal organizational barriers
- lack of regulatory requirements for travel plans
- personal taxation and commuting issues
- the poor quality of alternatives in the UK (particularly public transport)
- lack of experience due to the novelty of the concept.

It is clear that travel plans are a different sort of response to managing the transport impacts of an organization from traditional solutions. Building car parks is, organizationally, a relatively simple task. It is undertaken by one self-contained department (usually estate management), involving generic civil engineering skills that are to be found in that department already. Travel plans are very different and require skills that are usually not present in many organizations. They can also be difficult to justify, as a travel plan does not fit comfortably into many organizational structures.

There is potential in the travel plan concept. If travel plans are well designed and implemented consistently, they can reduce single-driver car use by at least 10–20% and possibly more. The target for modal shift in the index model is 23% – so the best travel plans are approaching what is needed (although a similar success rate would be required from other mobility management measures as well).

However, such travel plans are rare. There is a need to extend good practice and to integrate travel plans with other behavioural change measures, such as general investment in public transport, tax changes to reward 'green' travel, and measures to cut the distances we need to travel. In addition, of course, more general travel behaviour measures that address individuals are necessary, in order to hit the target in the Chapter 1 index model. Nevertheless, it is clear from the evidence here that achieving such a target is possible. Whether institutions, government and individuals consider the environmental and congestion costs of our transport problems to be sufficiently serious to merit such actions is another matter.

The technological and mobility management approaches share common features. They both have the potential to contribute significantly to achieving sustainable transport, but neither approach is adopted sufficiently widely to do so. The key challenge is to expand the use of low carbon fuels, high fuel economy vehicles, travel plans and other mobility management measures and bring them into the mainstream.

Figure 4.9 Do we have a real desire to leave this behind?

References

Cheshire County Council Travelwise (2002) *Commuter Plans in Cheshire: Steps to success*, Travelwise Team, Cheshire County Council, http://www.cheshire.gov.uk/travelwise/home.htm [Accessed 22 May 2003]

Collins, A. (undated) 'Stepping Hill NHS Trust Sustainable Transport Plan', Powerpoint presentation, Stepping Hill NHS Trust.

Department of the Environment, Transport and the Regions (1999) *Preparing your Organisation for the Future: The Benefits of Green Transport Plans*, London, The Stationery Office.

Department for Transport (2002a) *Making Travel Plans Work: Case Study Summaries*, London, The Stationery Office.

Department for Transport (2002b) *Making Travel Plans Work: Lessons from UK Case Studies*, London, The Stationery Office.

Department for Transport, Local Government and the Regions (2001a) *National Travel Survey 1998–1999*, London, The Stationery Office.

Department for Transport, Local Government and the Regions (2001b) *Evaluation of Government Departments' Travel Plans*, Report for the DTLR, London (unpublished).

Energy Efficiency Best Practice Programme (2002) *A Travel Plan Resource Pack for Employers*, Energy Efficiency Best Practice Programme, London, The Stationery Office.

Eyre, N., Fergusson, M. and Mills, R (2002) *Fuelling Road Transport: Implications for Energy Policy*, London, Energy Savings Trust.

Future Energy Solutions (2002) *Heathrow Airport Retail Consolidation Centre*, Report GPCS402, Energy Saving Trust, http://www.transportenergy.org.uk/bestpractice/ [Accessed 21 May 2003]

NHS Estates (2001) *Sustainable Development in the NHS*, London, The Stationery Office.

Royal Commission on Environmental Pollution (1994) *Transport and the Environment*, London, HMSO.

Rye, T. (2002) 'Travel plans: do they work?', *Transport Policy*, Vol. 9, No. 4, October, pp. 287–98.

Transit (2002) 'First offers Travel Plans to Hants businesses', No.191, 6 September, p. 32.

Chapter 5

Government Policies for Managing Energy Use

by Horace Herring

5.1 Introduction

This Chapter looks at the role of energy efficiency in national energy policy and, in particular, the debate over the extent to which promoting energy efficiency can reduce national energy consumption, and hence lower greenhouse gas emissions. It starts, in Section 1.2, by locating this debate in the context of our consumer society that exhorts us, in many aspects of our lives, to be more efficient, to save money and to cut costs. Amongst these exhortations are advertisements calling for greater efficiency in energy use, but these tend to be diluted by the overall volume of general advertising. In any event, saving energy is a subject of low importance to most people.

Section 1.3 then looks at the so-called 'market barriers' that are widely believed to prevent consumers investing in energy efficiency, even though it is in their economic interest to do so. This topic is a matter of considerable debate in academic and policy-making circles. The overcoming of such 'barriers' is held to be a legitimate goal of the current UK Government.

Section 1.4 examines the justifications for energy efficiency policies over the last three decades, which have rested on many non-energy arguments, such as promoting 'economic modernization' and technical change, improving the housing stock and preventing 'fuel poverty', and promoting price competition. It briefly discusses the policies and actions of UK Government during this period, paying most attention to the Home Energy Efficiency Scheme (HEES).

In Section 1.5 we then examine the current UK Government policy of promoting energy efficiency in order to reduce national carbon dioxide emissions. It looks in detail at the sequence of assumptions that underlie this policy, the existence of a large potential for energy savings and the means of assessing this potential. It also examines the links between energy and economic growth, and the difficulty in making international comparisons.

Section 1.6 looks at the 'rebound effect': the tendency for consumption to increase due to a lowering of the (implicit) price of energy bought about by efficiency improvements. Whether this increase in energy consumption outweighs the energy savings is much contested; as with all macroeconomic effects, the causal links and the magnitudes of the impacts are very difficult to prove. Historically, the evidence is that economic growth is linked to increased energy consumption - under the conditions desired by the UK Government of continued economic growth and low energy prices.

In Section, 1.7 we end with a critique of energy efficiency in our consumer society: is it just a means to increased (albeit 'greener') consumerism? We look at the historical and political roots of efficiency arguments dating back to the 'conservation era' in early century USA, which still influence industrial and governmental thinking. Then we outline some of the ethical arguments about the moral virtue of conservation (consuming less) and the need for 'sufficiency'– views adopted by many environmental economists, from Kenneth Boulding and Herman Daly onwards (see for example Boulding, 1966 and Daly, 1989). This poses the question of whether, ultimately, a sustainable society requires the end of economic growth. What policies would be required to 'de-link' energy growth from economic growth?

BOX 5.1 Definitions: Energy efficiency, energy conservation, energy intensity, factor four and factor ten

The concept of 'energy services' – services that energy uniquely can provide, such as heat, light or motive power - was introduced in Section 1.5 of Book 1 *Energy Systems and Sustainability.*

Energy savings bought about by a reduction in the consumption of energy services are usually considered to be due to 'energy conservation'; while energy savings achieved without a reduction in energy services are considered to be due to 'energy efficiency'. As the Royal Commission on Environmental Pollution (RCEP) remarks 'energy conservation implies reductions in the consumption of energy services. That could be achieved simply by making do with less energy by turning thermostats down and tolerating lower temperatures, for instance' (RCEP, 2000, p. 85) Energy efficiency, however, involves 'obtaining more useful heat, light or work from each unit of energy supplied, either as a result of technological improvements or by reducing waste; in other words, obtaining the same services with less use of energy' (RCEP, 2000, p. 85).

The concept of energy efficiency can only be meaningfully applied to a specific piece of equipment (like a boiler or an engine) where energy inputs and (energy service) outputs can be directly measured. For more complex systems, like a refrigerator, a house, a car, or a factory the term *energy intensity* is used, which can be defined as the ratio of energy use to a relevant indicator of physical activity or economic output. The relevant indicator varies by sector. For industry it might be tonnes per unit of output, or of value added. For the domestic sector it might be energy per unit of heated floor space, or in the transport sector, energy per vehicle- or passenger-kilometre. Terms that are related to energy intensity include *specific energy consumption* (SEC) and *unit consumption*, for example kWh per year used to quantify the annual energy consumption of a refrigerator of a given size and features.

Energy efficiency is the inverse of energy intensity or specific energy consumption (SEC); a doubling of efficiency implies a halving of energy intensity.

'Factor Four' is a phrase coined by advocates of radical energy efficiency improvements to describe the possibility of doubling the wealth of a society but halving its energy use - that is, achieving a four-fold increase in energy efficiency, which implies a 75% reduction in energy intensity (von Weizsacker *et al.*, 1997). A more extreme version of this idea is 'Factor Ten', that is achieving a ten-fold increase in energy efficiency (or a 90% reduction in energy intensity).

5.2 Consumers and efficiency

Consumers in today's society are bombarded by advertisements on how to save money and cut costs. TV commercials on cheaper motor insurance are just a phone call away; get interest on your current account through internet banking; collect coupons in the paper to get special offers at the supermarket. Consumer groups and government agencies do calculations that reveal how, nationally, consumers could save hundreds of millions of pounds if they took advantage of the best offers.

Amid all this advertising to be more efficient, to cut your costs and to save money are advertisements for another product: energy efficiency. Again we are informed of the large savings we could make with just a little effort, and organizations like the Energy Savings Trust assert that the UK could save many hundreds of millions of pounds (and avoid the emission of millions of tonnes of CO_2) if only consumers used energy more efficiently.

Most consumers pay little heed to the vast majority of these ads. However some of them are very successful, particularly those that only involve a one-off effort, such as purchasing car insurance by phone or ordering goods over the internet. Those that involve (or are perceived to involve) more substantial effort, like changing bank accounts, are less successful. Those that may involve a change in behaviour, like drinking less, driving slower or eating better, are the least successful.

The reasons why consumers do not respond to measures that are in their economic interest are a subject of deep concern to policymakers, a field of study for academics and a cause of anguish to advertisers. A superficial response is to blame the consumers for being irrational, lazy, uninterested in saving money, apathetic, dumb etc. While individual consumers may be any of these things, it is a great mistake to believe that any *group* of consumers, like the old, the young, the unemployed, single parents, women etc has these characteristics. If people do not heed our message, the problem probably lies with the message, not the people. Thus when consumers do not respond to the energy efficiency message and behave (in the experts' view) 'economically irrationally' the answer is not to question their rationality but to understand their behaviour.

5.3 Market barriers

People are saturated with ads every day to buy this or do that. Even if they agree with the message – and few dispute the merits of cutting costs and saving money - their time is limited and they have to set priorities for action. It is basically the 'hassle factor' - how much effort is involved in changing banks, getting a quote by phone, buying an energy efficient light bulb? People subject their effort to a crude cost-benefit analysis: what will I save for this effort, is it worth the hassle? Academics refer to this hassle as the 'transaction costs' and often these present a formidable obstacle to achieving action.

Is five minutes on the phone worth saving £50 on motor insurance, is fifteen minutes on the web worth saving £5 on a book, is an hour spent shopping in the high street worth saving £10 a year on an energy efficient fridge? The easier it is made to save the more people will respond, hence the growth in selling services like insurance, travel and banking, over the phone or web. It avoids the chore of visiting shops. It also, most crucially, saves consumers' time (except when you are put on hold or web connections are slow!).

Consumers can also use the phone (and web) to save money on their energy bills by switching to a supplier who may give them a lower price and bundle all their services (electricity, gas, phone) together, which can be convenient. However while this may save money it is unlikely to save energy, and may even increase it since consumers may be tempted to increase their comfort levels as their energy now cost less - the so called 'rebound effect' (see Section 5.6 below). However, despite over a decade of competition amongst suppliers and the ease of doing so, only 40% of consumers have switched supplier (NAO, 2003).

There are also telephone-based services (and web sites) that give advice on saving energy and how to be more energy efficient. These are often run by local Energy Advice Centres and have had some success. However their advice is general and falls into the following consumer-perceived categories:

1 the 'common-sense' - switch off lights that are not being used, turn the thermostat down, draw the curtains in winter.

2 Measures involving some change in behaviour: e.g. showers instead of baths, driving more slowly

3 Measures requiring further action – e.g. visiting shops, getting quotations

4 Measures that are not applicable to your circumstances e.g. you don't own your home.

It should be no surprise that consumers fail to take advantage of many of the offers to save money. Firstly, they may not believe the offers; secondly, they may think they are not relevant to them; thirdly, they just haven't got the time to make the effort - and finally, there are more fun things to do in life that worry about saving £10, or even £50, a year.

Government concern

However one organization that does worry about it is the Government. It believes that market competition, economic productivity and getting something for less are signs of efficiency and of competitive markets working to drive down costs. Consumers failing to take advantage of clearly economic options are thus considered by some energy analysts as a sign of 'market failure' and the existence of 'market barriers'. There is, however, a wide-ranging debate over the exact nature of such 'market failures', over what are the 'barriers' to achieving a more economically efficient outcome, and over what actions a Government should take to remove 'market imperfections'.

Nevertheless policy makers promoting energy efficiency (including the present UK Government) believe there are substantial market barriers that inhibit people from investing in it. There is now an extensive academic literature on the nature of these barriers, drawn from a wide variety of disciplines: institutional economics, management science, social psychology, sociology and political science (for more details see Chapman and Eyre, 2001).

According to a 2001 report from the Performance Innovation Unit (PIU, now called the Prime Minister's Strategy Unit), based at the UK Cabinet Office, there is a broad consensus among energy efficiency analysts and professionals on what constitutes the key problem. It is that:

> Energy users, in households and most businesses, do not seek to optimize the economic efficiency with which they use energy. In a complex world, people have many concerns and most have a higher priority than energy efficiency. So projects that are primarily about energy efficiency are often not even considered. And in investment decisions and purchases that involve energy use, energy efficiency is usually a minor consideration.
>
> (PIU, 2001, para 3.9).

Box 5.2 below list the main barriers identified by the PIU Report.

The PIU Report notes (3.10) that to the vast majority of business and households, energy bills are a minor concern, averaging 3% of expenditure in households and only 1.3% in business. The only exceptions where energy costs are of importance are a few energy-intensive industrial users (steel, cement, chemicals etc) and households facing 'fuel poverty' (see Section 5.4.3)

BOX 5.2 Market failure and the barriers to energy efficiency

The barriers to investment in improved energy efficiency may be described in a variety of ways, largely depending on the disciplinary background of the analyst. In neo-classical economic theory, many are market failures. The following may be identified.

- Most energy consumers have very imperfect information about energy efficiency opportunities and, especially in the domestic sector, may distrust information from vested interests.
- Current market structures often require energy users to expend time and money to gather and assess information that is already available to suppliers.
- Better information on capital costs than that for running costs leads to adverse selection of inefficient goods.
- Capital markets are incomplete for many borrowers, for example low-income households cannot borrow even for very cost effective projects.
- Inadequate contractual relationships with builders and other traders result in 'moral hazard', in sub-optimal specifications and the risk that that projects may not be implemented correctly.
- Tenancy contracts and rental values provide split incentives for energy efficiency investment in rented properties.
- Household investment in energy efficiency has less beneficial fiscal treatment than corporate investment in energy supply, as taxation is based on income rather than profits.
- Regulatory structures preclude potentially beneficial long-term contracts between licensed energy suppliers and domestic customers.
- The price of energy, in most cases, fails to take into account the environmental costs associated with its supply and use, i.e. there are externalities.

Source: PIU, 2001

The tendency of consumers to restrict their attention to only a number of issues is known in the academic literature as 'bounded rationality' (a concept first developed by economist Herbert Simon in his 1960 book *The New Science of Management Decision*). Consumers (like bureaucrats with an overflowing in-tray) have many issues competing for their attention, and it is to be expected that only the most important to them will be dealt with. The rest are put in the mental 'pending' tray or, if too troublesome, deliberately forgotten about.

The priorities of energy consumers are not the same as the priorities of energy efficiency agencies, and as the PIU Report admits (3.11):

> It is clearly impractical for Government to seek to re-order the priorities of every energy user. The principal barrier to energy efficiency is therefore not easily open to 'correction'. Making energy use markets 'perfect' is therefore not a sensible goal. But this is not a case for inaction. The economic inefficiencies are large and a large number of policy interventions may improve them, even if they do not remove them. In the language of neo-classical economic theory, the policy options are 'second best', but they can still be a great deal better than no action at all.

However not all would agree. Some economists who dispute the existence of 'market barriers' would argue that any Government action will only worsen market imperfections. Furthermore it is argued that the imperfections in the energy efficiency market are no worse than exist in other markets, such as insurance or food, and that we no more need an Energy Efficiency Office than we need a Food Efficiency Office or a Travel Insurance Efficiency Agency.

This debate about the existence of market barriers and the need for Government action is a complex one, often driven by political ideologies over the role of markets and Government regulation (see for example Wirl, 1997; Sorrell *et al.*, 2000). The main question is: what should the government do about what it perceives as the 'economic irrationality' of consumers? Should there be education campaigns to persuade them of the error of their ways (preaching), or should they be forced to change their minds through regulation and higher prices (the stick) or bribed through incentives and subsidies (the carrot)? Should we be exhorted to be more efficient through moral persuasion (it is good for us and good for the planet) or persuaded through economic self-interest (we can save money and spend it on luxuries)?

In the next section we examine the current UK Government's policies on energy efficiency. Obviously to it, and to agencies promoting energy efficiency, these policies are very important, realistic and achievable. However when reading about them think about your own attitudes to saving money - how much effort would you make to save £50 or £10 a year. Do the policies, as the Government hopes, 'work with the grain' of real decision-making?

5.4 Historic government policies on energy efficiency

Over the last three decades the UK Government has explicitly promoted energy efficiency (initially, it used the term 'energy conservation') with the aim of lowering the rate of growth of national energy use. However this desire for energy conservation (lower energy use than otherwise) has had to coexist with other policy aims that would tend to increase energy consumption, such as low energy prices and continued economic growth. The end result has been that economic growth has outpaced efficiency improvements, and UK national energy consumption has continued to increase (see Book 1, Chapter 2, Figure 2.8), albeit slowly.

Reasons for promoting energy efficiency

So does this mean that the UK Government should not promote energy efficiency? No, not at all; there are many valid and proven reasons to promote energy efficiency, as successive Governments have done since the 1970s, namely:

(1) to encourage economic productivity and growth, and spur technical change, i.e. to encourage 'economic modernization'

(2) to create export industries

(3) to help consumers adapt to higher energy costs

(4) to improve the housing stock and eradicate 'fuel poverty'

(5) to encourage fuel competition and keep down energy prices

(6) to reduce CO_2 emissions by reducing (fossil fuel) energy use.

(1) Economic modernization

The first of these rests on successive Governments' belief that they have a role to play in 'economic modernization' and sponsoring research into new technologies As the PIU Report comments (PIU, 2001, para 4.3):

> 'Most of the potential for future advance in energy efficiency will rely on technological improvements in materials technology, design and control. These are the types of change where the pace may accelerate in the knowledge economy. Energy efficiency is therefore expected to be an integral part of economic modernisation.'

This is a long established policy, and has been implemented by specialist economic and technical agencies. Work on funding and disseminating the latest advances in energy efficiency was first carried out by the fuel industries (1950–70s) when they were nationalized, then later in the 1970s by a specially created agency ETSU (the Energy Technology Support Unit), then in the 1980s by the EEO (Energy Efficiency Office), followed in the 1990s by the EST (Energy Savings Trust) and other 'quangos' (Quasi-Autonomous Non-Governmental Organizations) such as the Carbon Trust. Other research on energy efficiency is conducted by manufacturers of domestic appliances and cars, who play an important role in the setting of the EU's voluntary and mandatory labelling and standards schemes.

(2) Export industries

Some countries, such as Denmark, appear to have been more effective than Britain at creating exports for energy efficient products. The creation of successful export industries usually relies on a large domestic market, which can be stimulated by the imposition of national standards. In the UK, successive governments have however been more concerned with protecting UK manufacturers than promoting efficient products. This may lie behind their reluctance to support mandatory appliance standards, which would disadvantage UK manufacturers.

(3) To help consumers adapt to higher energy costs

Since the 'energy crisis' of 1973–4 successive governments have used various techniques to spread the energy conservation and efficiency message, initially to mitigate the impact of higher energy prices on consumers. These are:

(a) Exhortation, involving publicity and awareness campaigns,

(b) Information: specific targeting of the message,

(c) Regulation: setting minimum standards,

(d) Taxation,

(e) Incentives: giving grants or loans for investment,

(a) Exhortation.

This involves press or TV advertising to a mass audience, generally domestic consumers, to create awareness of the problem with some very general advice and information on where to find further help. An early example was the 'Save It' campaign in the mid 1970s run by the UK government. More targeted advertising, such as posters and leaflets, can then reinforce the message and provide sources of basic advice to the general public on ways to reduce energy use in the home, office or factory. As with all advertising, the message must resonate with consumers if it is to be successful. This originally happened with the conservation message of 'Save It', but with the decline in the urgency of the 'energy crisis' of 1973–4, consumer attitudes changed and they became hostile or unreceptive to the message of conservation (see Section 5.1.4).

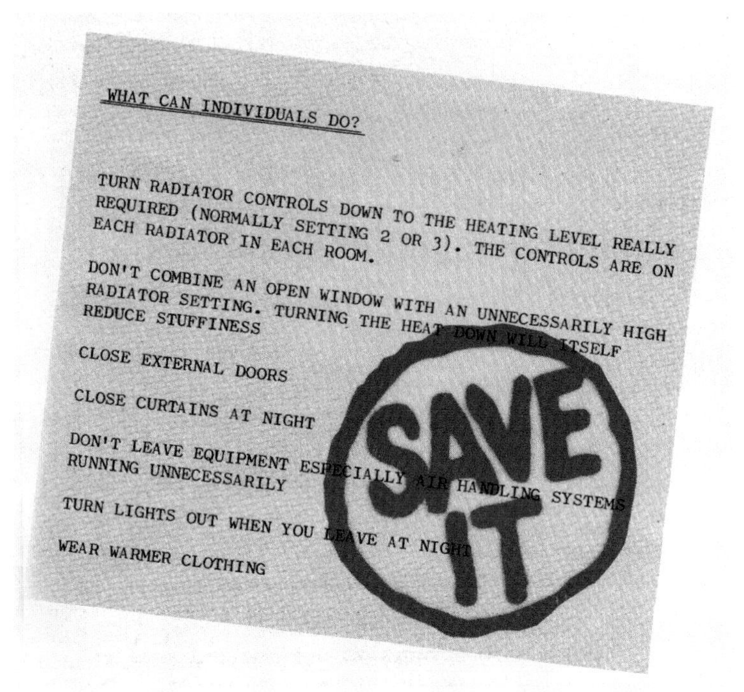

Figure 5.1 'Save it' leaflet produced by UK government

(b) Information

This involves targeting of the message, for example through energy information centres or telephone advice lines to answer specific questions from consumers. Another option is energy labelling of appliances and homes so that people can compare efficiencies and see what the running costs are likely to be. This has been done in the EU since 1994 on domestic appliances, first with fridges and freezers then extending to other household goods (washing machines, dryers, dishwashers etc). In the UK energy labels have been required for all new buildings since 1992, but this labelling scheme has not been extended to existing homes (when they are offered for sale), although this has been done in Denmark.

(c) Regulation

This is a long-standing way for governments to ensure minimum standards of energy performance. These can either be voluntary (agreed by manufacturers) or compulsory (set by law). In the UK, Building Regulations have required certain levels of insulation in new homes since 1965, and these have progressively been upgraded, with the latest being in 2002. However by Continental European standards UK insulation levels are low.

In the USA there have been compulsory energy standards for domestic appliances since 1989 and voluntary standards for office equipment (the Energy Star Office Products program) from 1992. In the EU, despite many years of debate, there are only voluntary standards, with opposition coming from certain appliance manufacturers selling low-efficiency goods, who are afraid of losing sales, and from some governments opposed to compulsory regulation.

(d) Taxation

Energy taxation has proved unpopular with UK consumers and many governments are currently committed to 'low taxation' policies. In the UK, while petrol and diesel for vehicles has always been heavily taxed, domestic fuels only have a reduced 5% VAT rate. Attempts by the Conservative Government in 1994 to raise the rate from 8% to the standard 17.5% proved very unpopular and were defeated (the subsequent Labour administration then reduced it to 5%). Similarly, massive protests in September 2000 by road hauliers about high fuel prices (see Book 1 Figure 1.2), perceived to be due to increasing fuel taxes, resulted in the UK Government abandoning its policy of annually raising fuel taxes.

However some countries such as Denmark and Norway have used energy taxation for many years in order to reduce dependence on imports and more recently have introduced carbon taxes to reduce CO_2 emissions and promote renewables. For the last decade the EU has been considering the introduction of an EU-wide carbon or energy tax in order to reduce greenhouse gas emissions. This proposal has run into strong opposition from industry -worried about the effect on international competitiveness - and some governments (the UK included) politically opposed to new taxes.

(e) Incentives

These take the form of grants or low- or no-interest loans towards energy saving measures or toward the cost of individualized energy advice (energy audits). For households, the most common measures are financial support for insulation schemes: loft insulation, draught-proofing, cavity wall insulation and upgrades to heating systems and controls. In the UK the government has been reluctant to subsidize the investments of the majority of consumers, preferring to reserve its limited financial support for low-income consumers. The only universal grant was for loft insulation, available from 1978 to 1988 for all households and until 1990 for low-income households. From 1990, Home Energy Efficiency Scheme (HEES) grants for loft insulation and draught-proofing were only given to low income or pensioner households, but there have been limited subsidies (or rebates) totalling a few million pounds for the purchase of energy efficient compact fluorescent lights, refrigerators, and condensing boilers. From June 2000, funding for HEES has been expanded, and it is now marketed as 'Warm Front'.

(4) Improve housing stock

The UK has some of Europe's worst housing conditions, hence the inability of many poor households to afford high heating standards. Energy efficiency improvements have been promoted as part of a package to upgrade the housing stock. The rate of improvement in insulation levels in the UK housing stock is very slow, due to the very small (1% or less) rate of new build, with its much higher levels of insulation. Thus to accelerate the improvement in overall insulation levels it is important to concentrate on existing buildings. However insulation grants have been minimal (see Section 5.4.3 on HEES below)

Figure 5.2 Designed and built by British Gas, the Gaswarm Windsor house in Milton Keynes provides gas central heating, double glazing and fibre insulation to roof, cavity and underfloor

(5) Encourage fuel competition

Promoting lower energy prices has been the goal of all UK governments. First there was political control of energy prices by the nationalized industries, then after privatization the use of regulation and encouragement of competition amongst the privatized utilities. Fuel competition between electricity and gas utilities for the space and water heating market has also spurred efficiency improvements in new buildings and in process heating equipment. However this competition was on the basis of cost, and the lowest cost fuel (such as off-peak electricity) may not produce the lowest (primary) energy consumption and CO_2 emissions.

Government campaigns

Whilst the rationale for promoting efficiency remains constant, UK Government campaigns on energy efficiency and conservation have changed over the past three decades, reflecting their immediate political concerns. Since the 1970s these campaigns have fallen into three categories:

(1) To save energy: security of supply and depletion concerns

(2) To save money: promoting economic efficiency and competitiveness

(3) To save the earth: environmental concerns, especially global warming

The first campaign was undermined by increased fossil fuel production, assisted by enhanced oil and gas production techniques, the second by the decline in oil prices from 1986, but the third (and latest) is still strongly supported by the UK and many other governments.

Save it!

The first campaign in the 1970s to save energy was motivated by Government concern over security of oil and coal supply, which was threatened by the OPEC oil embargo and strikes by UK coal miners - see Box 5.3. There was also a desire to conserve national energy resources for the future, when it was forecast that energy prices would be much higher. Thus the emphasis was on energy conservation, to stabilize, or even reduce, the absolute level of energy consumption.

The then Labour Government launched its conservation programme in January 1975 with the slogan 'Save It'. It included such memorable hints on energy saving as 'wear warmer clothing', 'close curtains at night', etc. The public, faced with the reality of power cuts and imminent petrol rationing, responded well. The message chimed well with many of the older generation who remembered similar calls during World War 2 when there was fuel rationing, and to some of the younger generation enamoured with the concept of 'self-sufficiency' and the need for alternative sources of energy. The 'energy crisis' quickly passed (the miners went back to work and the OPEC embargo ceased) but the image of energy conservation as that of voluntary restraint, simplicity and hardship persisted.

BOX 5.3 **Security of supply**

Historically the main reason for governments to promote energy savings was to conserve supplies at times of crisis. For example, the coal shortages experienced in much of Europe and Japan after the Second World War led to efforts by a number of countries to conserve stocks, often reducing supplies to households to ensure industry could get enough to maintain production. The other obvious example was the 1973 oil crisis, as a result of which many governments introduced a range of short-and long-term measures to promote better use of energy as a means of reducing dependence on imported oil.

During the 1970s, concerns were also first raised about the 'Limits to Growth' (Meadows *et al.*, 1972) and that the world might run out of fossil fuels in the foreseeable future. Few people seriously believe that world reserves of fossil fuels are likely to be exhausted in the foreseeable future but supplies of conventional oil could, in the decade after 2000, reach a peak and thereafter start to become much more expensive (see Book 1 Chapter 7, Oil and Gas).

Long-term security of supply is not a factor that many of today's energy customers will take into account. Energy suppliers may consider it to a limited extent, but they are likely to be more concerned about maximizing short-term returns on capital. This suggests that security of supply is an externality that is not accurately reflected in the market price for energy.

Source: Owen, 1999, pp. 17–18

Western governments have been more relaxed about these security of supply issues since the collapse of oil prices (apart from a few spikes) from 1986, despite their continued dependence on Middle East oil. For instance, in the 2003 Energy White Paper (DTI 2003 – see Section 4.44), there is little concern expressed about the UK having to rely (in the future) on gas imports from Russia or oil from the turbulent Middle East.

Figure 5.3 Monergy leaflet produced by UK government

Monergy

During the 1970s 'energy crisis' President Carter of the US appealed to the American nation to embrace 'conservation', calling it the 'moral equivalent of war' (Moezzi, 2000, p. 522). He was derided by Ronald Reagan, his rival candidate for the Presidency who won by a landslide in 1980, and instead promoted energy production not conservation.

Similarly in the UK Margaret Thatcher won the 1979 general election and emphasized energy production, particularly nuclear power. Policy makers quickly dropped the concept of 'conservation' and instead called for 'efficiency', which fitted in well with their free market ideology of promoting economic growth through harnessing the greater efficiency of the private sector. Thus the second UK government campaign was based on saving money not energy. So out went the 'Save It' campaign and in came the 'Monergy' campaign which stressed that energy efficiency was good for businesses, and boosted economic productivity.

Saving the earth

The third and latest campaign, which has dominated energy efficiency policy since the 1990s, seeks to reduce the environmental impact of energy production and consumption. Since the late 1980s global environmental concerns have risen up the political agenda, first with 'acid rain', then the 'ozone hole' and finally 'global warming'. One way to reduce CO_2 emissions is to reduce fossil fuel use, and energy efficiency has been increasingly promoted as the key solution (or technology) to reduce energy (or more particularly fossil fuel) consumption.

However it is important to remember that global carbon emissions are not the only environmental impact from energy production and consumption: there are other important impacts at the local, national, and regional level (see Book 1, Sections 13.5–13.6). These impacts are very varied and can range from local health concerns over childhood asthma due to motorway traffic to national outrage at forest damage due to the sulphur emissions of a nearby country. A whole range of issues, like child health, road safety and urban planning, are increasingly been seen as environmental issues, affected by the way our society produces and consumes energy. Thus energy efficiency, in that it affects the level of energy use, is seen as an important aspect of environmental policy.

The home energy efficiency scheme

This ability of energy efficiency to entwine itself with other non-energy policies is clearly seen in the 'fuel poverty' debate, where it is promoted as the solution to enhancing heating standards amongst the 'fuel poor' - those households that would need to spend more than 10% of their income to achieve an 'adequate standard of warmth'. This problem can be seen in a variety of ways: as a social welfare issue (caused by inadequate incomes), a housing issue (caused by poor quality housing) or an energy issue (poor insulation levels). The solutions are as varied as the causes: raising low incomes, insulating old houses, building new houses, or reducing energy prices.

The rather limited responses of past governments to this problem have been motivated more by Government ideologies rather than a clear social, fiscal or energy strategy. During the 1980s the Department of Energy provided only limited financial support for low income households via home insulation schemes run by local organizations and managed nationally by Neighbourhood Energy Action (NEA). In 1991, in response to the campaigns of the fuel poverty lobbyists, the Home Energy Efficiency Scheme (HEES) was created with an annual budget of £26 million and again managed by NEA. Funding for HEES was increased four-fold in 1994 to cover all those aged over 60, when the government introduced its politically-sensitive proposal to raise the VAT rate on domestic energy from 8% to 17.5 % (in the end the proposal was defeated but the HEES funding increase remained).

The current Government's goal is to eliminate fuel poverty by 2010 in the three million 'vulnerable' households that suffer from it. This is to be achieved through energy efficiency schemes (it re-launched HEES in 2000 with increased funding) and lower energy prices (see also Book 1, Chapter 12 and DEFRA, 1999)

However while HEES may be an effective energy efficiency scheme it is a poor energy conservation one. Research has found that, in low income households, about half of the saving is taken up in improved comfort (fewer draughts and higher temperatures), as previously most such households could not afford to heat their homes adequately (Milne and Boardman, 2000). As Gill Owen, one of the co-founders of NEA, remarked that:

> Most low-income households take up most of the benefit of insulation measures in terms of comfort rather than through any marked reduction in energy consumption. A government with a strong energy conservation objective might therefore feel that resources would be better targeted at those sectors (better-off households and businesses, for example) which would be able to save the most energy.
>
> Owen, 1999, p. 101

This very valid point, that it is the 'fuel rich' not the 'fuel poor' who can produce the most energy savings - that is, achieve the most energy conservation - illustrates the complexities and contradictions resulting from entwining energy efficiency with other non-energy policies. Measures that encourage conservation, such as increased energy prices bought about by a higher rate of VAT on fuel, work against other policies such as alleviation of 'fuel poverty'. It is very hard to design energy policies that are fully compatible with other aspects of government policy. We all want, as the government declares in its latest Energy White Paper (DTI, 2003), 'secure, reliable and affordable energy from sustainable and diverse sources' or, in a nutshell, to have 'cheap energy with no adverse consequences'. The extent to which this is a feasible goal is debatable, as are the technical means to achieve it.

5.5 Current government policies

The current UK Government, as it states in its 2003 Energy White Paper (DTI, 2003), believes that:

1. there is a large potential for improving the efficiency of energy use,
2. this could produce big economic savings to consumers,
3. future national fossil fuel use could be reduced and hence CO_2 savings made,
4. there are market barriers that prevent consumers achieving savings,
5. the way forward is Government action to overcome these barriers.

These claims are examined below. The first two rely on economic and technical analysis of the patterns of national energy use, i.e., how consumers use energy. Here the potential relies on the fact that new equipment is a lot more efficient that existing equipment; dispute occurs mainly over the magnitude of the potential and the speed with which it could be achieved. The third point depends on the 'rebound effect' - basically the decisions by consumers on how they will spend the money saved; dispute centres on the macroeconomic effects of these decisions over a long period of time. The fourth point has some economic validity but the magnitude and relevance of such market barriers is debatable; while the last point involves political choices by Government.

Potential for energy efficiency

New equipment and products are generally more energy efficient than older ones. This is all part of the trend of continual product improvement, whereby each new generation of equipment, by using the latest technologies, can provide the same energy service using less materials and energy. However products often incorporate new features that provide a new range of services, and these often negate the energy savings. For instance, cars with air conditioning, frost-free fridges, TVs that can be switched-on instantly from 'standby', or larger homes with more bathrooms. Sometimes there is a radical shift in the technology which result in products that both have significant energy savings and better performance: compact fluorescent versus incandescent bulbs; microwave versus stove-top cooking; laptops versus desk-top computers; e-mail versus postal mail.

However it must be borne in mind that new technologies have different characteristics to older ones and not all consumers will find it convenient or desirable to use them, despite the obvious energy savings. For instance, some consumers prefer the soft yellow light of incandescent bulbs to the white light from fluorescents, meals baked in the oven taste different from those heated in the microwave; email doesn't have the romance of the letter; the open coal or wood fire is more cosy than gas central heating.

There is a long history of some people rejecting modern technologies in favour of older ones, some because they can't afford the new ones (e.g. the rural poor), others because they prefer an old-fashioned lifestyle (as in the fashion for Edwardian country houses), and others who reject modern technology for political, ethical and religious reasons (such as hippies, or the Amish sect in the USA). Technologies are often perceived as symbols of the economic and industrial structures that have produced them: for example, people are far more likely to accept and use free compact fluorescent lamps (CFLs) given out by a local community group which they support than those offered by a remote utility with which they may have an antagonistic relationship. Thus it must be remembered that products and technologies are seldom 'value-free'; the context in which they are promoted and distributed can crucially affect their uptake (Bijker, 1995).

Assessing energy saving potential

Calculating the potential for energy saving is superficially straightforward but dependent on many assumptions. For instance a 75 W incandescent bulb can be replaced with an 18 W CFL giving roughly the same light output. The energy consumption of the CFL will be 18/75 or 24% of the incandescent bulb; thus we can say we have a 76% saving. If we assume the light is on for a 1000 hours a year (nearly 3 hours a day) then annual saving from using the CFL is:

$(75 - 18) \times 1000 \times 0.001$ kWh, or 57 kWh.

If electricity cost 5p/kWh, then we have saved

$57 \times £0.05$ or £2.85 (for a similar calculation see Book 1, Box 12.3)

If we assumed that nearly every household in the country (say 20 million) replaced one incandescent bulb with a CFL bulb, and used it 1000 hours per year, then national savings per year would by 20 million times 57 kWh, or 1140 GWh (1 GWh = 1 million kWh) and be worth £57 million a year. If the generation of 1 kWh of electricity results in the emission of greenhouse gases equivalent to 0.13 kg of carbon (assuming the UK fuel mix for electricity generation in the year 2000, see Chapter 14, Figure 14.1) then each CFL saves 0.13×57 or 7.4 kg carbon per year; and nationally, for the 20 million homes, some 20 million $\times$ 7.4 $\times$ 0.001 (1 tonne = 1000 kg) or 0.148 million tonnes carbon will be saved.

Note:

To convert kg of carbon to kg of CO_2 multiple by 44/12 or 3.67; 0.13 kg carbon equivalent to $0.13 \times 3.67 = 0.477$ kg CO_2

Thus we can say that the potential for energy saving from installing one CFL in every home (under the assumptions we have made) is over £50 million a year or nearly 190 000 tonnes of carbon. More complex calculations can be made taking into account differences in bulb wattage, hours of use, and house type. By making these sorts of calculations the Performance and Innovation Unit (PIU) estimates that nationally 1.4 million tonnes of carbon could be saved from more energy-efficient lighting in homes. Similar calculations can be made for other energy efficiency measures, like condensing boilers, cavity wall insulation, and more efficient appliances. The measures are split by the PIU into those which are judged to be cost-effective to consumers (i.e. those giving an 'Internal rate of return' of over 15% - see Book 1, Chapter 12, p. 27) and those which are technically feasible but economically unattractive to consumers.

Table 5.1 shows the PIU estimates of the potential for energy savings and carbon emission reduction from the UK domestic sector. These 'best estimates' indicate that over a third of domestic energy use could be saved with measures that are currently economic to consumers, a figure that rises to over half if you include with measures that are technically feasible but not yet economic to consumers. It must be remembered that these are estimates, based on many assumptions, and it is highly unlikely that all these savings could be realised. Potential is rarely achieved. As the PIU report admits (2.8) 'The fraction that can be achieved by 2010 is constrained by stock turnover, installer capacity and householder attitudes'. Nevertheless the UK Government is hopeful that current and future programmes will achieve annual savings of between 4–6 million tonnes of carbon (MtC) by 2020 or one-eighth of current domestic emissions of 40 MtC (see DTI, 2003, Table 2.1).

Table 5.1 Annual energy savings and emissions reduction from domestic energy efficiency measures

Measure	Mtoe saved	MtC saved	Internal Rate of Return
Economic Potential			
Loft insulation	1.7	1.4	16%
Cavity wall insulation	3.2	2.6	32%
Hot water cylinder insulation	0.4	0.3	200%
Condensing boilers	6.5	5.3	27%
Energy efficient lighting	0.9	1.4	50%
Energy efficient appliances.	1.9	2.9	19%
Double glazing	2.1	1.7	19%
Heating controls	0.5	0.4	38%
Small-scale CHP	0.2	0.3	19%
Sub-Total (Economic potential)	17.4	16.3	
Percentage of domestic total	37%	41%	
Additional technical potential			
New district heating CHP	0.6	0.9	
Solid wall insulation	3.4	2.8	3%
Draught proofing	0.4	0.3	6%
Solar water heating	2.0	1.6	–8%
Ground source heat pumps	0.4	0.7	0%
High performance glazing	1.5	1.2	–2%
Sub-total	8.3	7.7	
Total	26.01	23.8	
Percentage of domestic total	54%	57%	

Source: PIU, 2001, Table 1

Similar analyses can be done for other sectors of the economy. The PIU estimates (para 2.10) that in the service sector (commercial and public buildings) the economic potential is 21%, with technical potential nearly double that at 39%. In industry the economic potential is 24%, with technical potential at 36%. Most of the potential lies in the energy intensive industries, like steel, cement, paper, chemicals and engineering. In transport the saving are one third. Overall, the PIU report (in its Table 4) estimates that nationally about 30% of energy could be saved, with consumers saving £12 billion a year. This is a very significant potential, but how much could be achieved is a matter of dispute.

Stock turnover

It is very likely that the UK will be more energy efficient in the future than it is now. This is due to the process of 'stock turnover' whereby old equipment is replaced (generally at the end of its life) by more modern equipment which is normally more efficient. However this efficiency is usually a by-product of the design; few appliances are purchased just because they are more energy efficient (though there are exceptions, for example industrial motors and some 'chest freezers'). Normally consumers face a limited choice of efficiencies in their purchases; for instance cars are offered in a wider variety of colours than efficiencies. However the latest model of a refrigerator is usually more efficient than previous versions.

There has been a continuous trend of improvements in energy efficiency in most energy services. Over a century, Factor 4 or Factor 10 improvements have been achieved for some services. For instance there has been a factor 10 improvement in lighting efficiency during the twentieth century as the early carbon filament bulbs have evolved (through technical progress and innovation) into the CFLs of today - see Section 9.3 in Book 1. Similar trends in efficiency have been found in many industrial processes where it is fairly easy to calculate specific energy consumption (SEC). Figure 5.4 shows the historical development over two centuries of the SEC of pig iron and aluminium production, and of nitrogen fixation - note that the vertical axis uses a logarithmic scale.

There have also been substantial improvements in the efficiency of domestic heating systems, bought about by changes in heating equipment and insulation improvements in the home (see also Book 1, Chapter 3). The

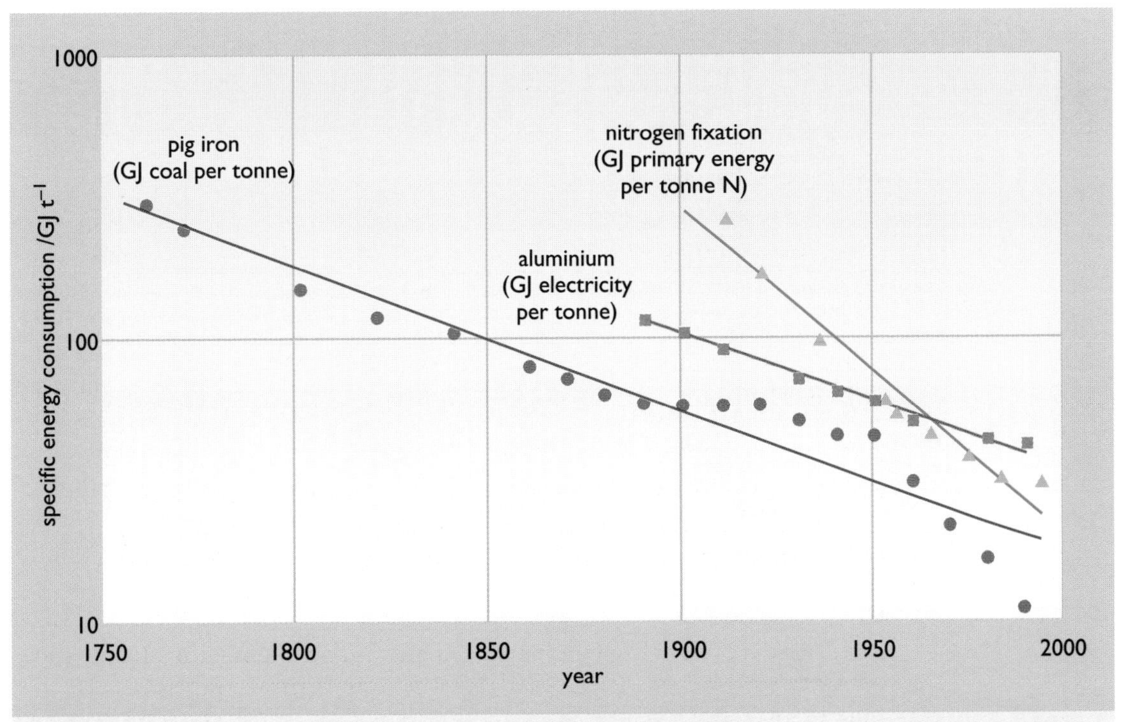

Figure 5.4 Specific energy consumption of pig iron production, aluminium production and industrial nitrogen fixation over the past two centuries (source: de Beer, 1998, Table 2.2)

open coal fire with an efficiency of about 25% has been replaced with modern gas boilers with efficiencies of around 70% and heat losses have been reduced by double glazing, cavity wall and loft insulation. Quantifying the exact magnitude of the improvement is difficult because of the problem of defining space heating energy intensity for home heating. But it is interesting that domestic energy use per person has remained roughly constant over the last century despite tremendous improvements in thermal comfort and appliance ownership.

Energy and economic growth

Since the late nineteenth century, UK national energy consumption has grown at a slower rate than economic output, and between 1880 and 1990 the UK energy intensity fell by a factor of 3 (see Figure 3.17 in Book 1). This seems to be a characteristic of mature economies, where most of the industrial infrastructure has been established and the economy shifts towards service industries. Newly-industrializing economies show an opposite trend: energy use rises faster than economic output, up to a certain level then falls, as in mature economies.

Despite economies becoming more energy efficient, energy consumption still usually rises with economic growth (thought the linkage is not 1:1). National energy consumption seldom falls except during periods of economic recession when there can be severe industrial restructuring. This occurred in the UK in the late 1970s (see Chapter 2, Figure 2.8) and also occurred in the former Soviet Union in the 1990s as its obsolete and highly inefficient heavy industries closed down. All economic growth involves the use of some energy; it is hard to think of any (economic) activity that does not involve some energy use (apart from food energy) - perhaps the nearest examples would be leisure activities that take place in the open air, like classical music concerts, sports or car boot sales!

Some authors have argued that modern economies can, and indeed are, restructuring towards low energy intensity service industries; a process they term 'dematerialization' (see for example Bernardini and Galli, 1993; Hawken *et al.*, 1999). This is indeed taking place as western economies are exporting their heavy industries to developing countries and concentrating on developing their service industries like finance, tourism and information technology. A country could indeed reduce its national energy consumption (as Russia has done) by closing down its heavy industry and showing what appears to be big improvements in energy efficiency. However a country's impact on the global environment depends on its consumption, not its production. As western economies move towards 'knowledge economies' and export their heavy industries to developing countries where environmental standards are lower, then overall global pollution can rise. As a World Resource Institute Report commented (WRI, 1997) 'The air over Northern cities might seem clearer but the world's atmosphere is much the worse for wear'.

International comparisons

It is fairly easy to draw up rankings of countries by their energy intensities; what is more difficult is determining the reasons for the differences and the implications for energy efficiency. The US uses over twice the energy per unit of output of Denmark (see Book 1, Table 3.5); does this indicate that the US is wasteful in its energy use? One can point to signs of a lavish lifestyle in the US: large cars with low mileage per gallon, extensive homes with air conditioning, massive refrigerators, frequent plane journeys and poor public transport. The fact that the US uses five times the world average primary energy consumption (see Table 2.3 in Book 1) is to some a sign of moral degeneracy and to others a source of pride. As US President Nixon said in 1973:

> There are only seven per cent of the people in the world living in the United States, and we use thirty percent of all energy. That isn't bad; that is good. That means that we are the richest, strongest people in the world, and that we have the highest standard of living in the world. That is why we need so much energy, and may it always be that way.
>
> Quoted in Hilgartner *et al.*, 1982, p. 181

But is lavish consumption a sign of energy *inefficiency*? Is large inherently inefficient? What should be compared to what? A small to a large car, or the US pickup to the British Range-Rover?

In other areas such as manufacturing the US compares well in its energy efficiency to other countries (IEA, 1997). Undoubtedly if US citizens lived in the same style as the Danes their energy consumption would be lower, but the same could be said if the Danes lived in the same style as Indians. Is energy intensity just a feature of external national characteristics such as per capita income (high in US), energy prices (low in US) and population density (low in US), rather than moral virtue or enthusiasm for energy efficiency? Such questions are explored in work, for example, by Lee Schipper and his colleagues (Schipper and Meyers, 1992).

Feasibility of UK government policies

The above observations concerning the difficulties of defining and measuring energy efficiency (on a national scale) can be summarized as follows:

- Most products are becoming more energy efficient due to a process of technical innovation.
- These products are slowly introduced into the economy by a process of 'stock turnover'.
- This does not mean they use less energy, as often we get products with more features (larger size, more power etc).
- Over time, developed economies usually grow faster than their energy consumption; hence energy intensities tend to decline.
- National economies may be more 'energy efficient' but may still use more energy in absolute terms (due to economic growth).
- The only ways in which economies have used less energy (i.e. an absolute decline) is for there to have been the painful combination of economic recession and industrial restructuring, and/or the export of energy intensive industries.

One far-reaching goal that has been proposed is to have both economic growth and an absolute decline in energy consumption; that is, the complete de-linking of energy and economic growth - the 'Factor 4' approach (von Weizsacker *et al.*, 1997). The UK Government believes that this 'decoupling' is possible through the vigorous promotion of energy efficiency policies, and points to the fact that there has only been a 15% rise in energy consumption between 1970 and 2000, while the gross domestic product (GDP) has doubled (DTI, 2003, Para 2.16). However the time period chosen has a big impact on the validity of such statements: if the year 1983 (the trough of the 1980s recession) is chosen, then UK energy use has risen by 20%, while GDP has increased by two-thirds.

However the current concern of the UK Government is with reducing CO_2 emissions rather than energy use, with energy efficiency being promoted as a means to achieve an absolute reduction in national carbon emissions, through an overall reduction in energy demand (DTI, 2003). The Government has been influenced in its choice of this policy by the many studies (such as by the RCEP Report *Energy: The Changing Climate* - see Box 1.4 in Book 1) that have argued that it is possible to achieve large reductions (of the order of 60%) in national carbon dioxide emissions, partly through reducing national energy consumption by between a third and a half, and partly through energy efficiency improvements.

Like all projections of the future, these scenarios suggesting lower national energy consumption are debatable but cannot be proven wrong, at least not until the date for which they are set arrives (2050 for the RCEP scenarios). However the scenarios rest on two key assumptions: firstly that there is a large energy efficiency potential (see earlier in this section), and secondly that if this potential were achieved it would persist – that is, none (or very little) would be lost to the 'rebound effect'.

5.6 The rebound effect

The 'rebound effect' (or 'take-back' effect) is the term used to describe the effect that the lower costs of energy services, due to increased energy efficiency, have on consumer behaviour, both individually and nationally. It is an observation based on economic theory and long-term historical studies, and as with most economic observations its magnitude is a matter of considerable dispute (Schipper, 2000).

As with our case of the 18W CFL replacing a 75W incandescent bulb, the energy saving should be 76%. However it seldom is. Consumers, realising that the light now cost less per hour to run, are often less concerned about switching it off - indeed they may intentionally leave it on all night. Thus they 'take back' some of the energy savings in the form of higher levels of energy service (more hours of light). This is particularly the case where past levels of energy services, such as heating, were considered inadequate, as with homes suffering from 'fuel poverty'. The energy savings from increased levels of insulation may then be spent on much higher heating standards: the consumer benefits by getting a warmer home for the same or lower cost than previously (Milne and Boardman, 2000).

Much evidence on the magnitude of the direct rebound effect comes from US transportation studies where there is good statistical data on miles

BOX 5.4 Types of 'rebound effect'

A paper by Lorna Greening and colleagues (Greening *et al.*, 2000) looked at the rebound effect and identified four categories of effects. They describe these as:

(1) direct rebound effects

(2) secondary (or 'income') effects on fuel use

(3) market clearing and quantity adjustments (especially in fuel markets), and

(4) transformational effects.

The *direct rebound (or price) effect* is the increased use of energy services caused by the reduction in their effective price due to greater efficiency. This works exactly as would the reduction in price of any commodity. Similarly, the *secondary* (or 'income') *effect*, is caused by increased expenditure on *other* goods and services, due to the reduction in the cost of energy services. The *market clearing effects* represent the result of myriad adjustments of supply and demand in all sectors of the economy due to a change in energy price. Finally, *transformational effects* are long-term effects in the economy due to changes in technology, consumer preferences and even social institutions, bought about by the substitution of energy for labour, capital or time.

travelled per vehicle and petrol consumption (Greene, *et al.*, 1999). The results indicate that the number of vehicle miles travelled will increase (or rebound) by between 10% and 30% as a result of a 10% improvement in fuel efficiency. Similar results are obtained for domestic energy services, such as space heating (Greening *et al.*, 2000). The impacts of the three other rebound effects (see Box 5.4) are much more difficult to quantify, and the few studies done provide inconclusive results. Attempts to estimate the overall magnitude of the rebound effect, using theoretical economic models based on neoclassical growth theory, have again proved inconclusive, with the results dependent on assumptions about the 'elasticity' (see below) of substitution of energy for other factors of production - the greater the elasticity the greater the rebound (Saunders, 2000)

Price elasticity

The effect of perceived lower costs on energy use is termed 'price elasticity' – the ratio of the percentage change in energy use to percentage change in energy price. So a 10% rise in energy use following a 50% fall in energy price is a ratio of

$$\frac{10\%}{-50\%} \text{ or } -0.2$$

Similarly, a 10% fall in use following a 50% rise in price is an elasticity of

$$\frac{-10\%}{+50\%} \text{ or } -0.2$$

Price elasticities vary by commodity and over time, depending on the ability of consumers to respond to price changes, either through changes in behaviour, substitution of alternatives or changes in technology. So a rise in petrol prices can result in the following actions: driving less (behavioural change), taking the bus (alternatives), or purchasing a car with better mpg (technology change). In the short term (a few years) consumers' ability to respond is often limited by their inability to change their behaviour (they have to commute to work by car), by lack of alternatives (no bus to work), or their lack of capital to (immediately) purchase a more efficient car (it is too costly to scrap their existing one). Thus price elasticity is low, generally in the range –0.1 to –0.3. However in the long term consumers' ability to respond to changes in prices is much greater: they can change where they live or work, public transport can be expanded, and more efficient cars purchased (when existing cars need to be replaced).

If energy prices fall after a period where consumers have responded with efficiency changes, the result is a substantial fall in overall energy costs and a boost to energy consumption. Thus the oil prices rises of the 1970s stimulated more efficient car engines, giving cars with high mpg; when oil prices fell in the late 1980s, car manufacturers could offer consumers more powerful and better equipped cars (luxury 4 wheel drive jeeps) with the same fuel costs but lower mpg (see also Block 3, Units 23–24).

Income effect

The second effect on consumers of lower energy costs is the 'income effect' – i.e., their ability to spend the monetary savings on other goods and services in the economy, which will contribute to economic growth. As with our light bulb example, the consumer now has £2.85 extra to spend (the saving on the energy bill) – assuming no change in their hours of use. What they will spend this 'discretionary income' on depends on their current income levels: those on low incomes will use it for 'basic' goods; those on higher incomes on 'luxury' services. However almost all goods and services involve some energy consumption in their production, distribution and consumption, though their energy intensity varies markedly.

It is important to remember that money circulates in the economy, and passes through many transactions. For example the money you give the greengrocer for carrots, the greengrocer spends on petrol for his delivery van, the garage uses to pay wages, the mechanic to pay the rent, the landlord to buy insurance, etc. Each monetary transaction involves some energy expenditure and contributes to economic growth. It makes little difference if you don't spend the money but save it in a bank, for the bank will lend it to someone else to spend. This question of what the monetary saving is spent on is crucial to achieving 'sustainable consumption' - that is, consumption which has low environmental impact, which implies having low material and energy content.

Technological change

Most probably the greatest effect (in the long term) of lower costs of energy services is on the direction and pace of technological change and innovation in the economy. New goods and services will be created to take account of the possibility of lower costs (see Box 5.4 below on impact of the tungsten filament bulb; also Book 1, Section 9.3). For instance the range of uses for electric lighting has expanded greatly. The big increases in lamp efficiency, which lowers running costs, have stimulated the market for security and outdoor lighting; also street lighting has expanded tremendously during the twentieth century (Herring, 2000).

Manufacturers continually seek efficiency improvements in order to lower consumer costs and create new mass markets. In conjunction with this, utilities seek efficiency improvements to lower the cost of their product: it is usually better to sell a lot at a low profit margin than a little at a high

BOX 5.5 Early twentieth century lighting

In the early 1900s new light bulbs with tungsten filaments replaced those with carbon filaments. These new bulbs only used a quarter to half the electricity of the older bulbs. The immediate result for the electricity utilities, then heavily dependent on electric lighting sales, was a sharp drop in revenues. Some utilities responded to this loss of income by raising tariffs, and increasing the power of the bulbs.

However the more far-sighted utilities realised that this efficiency revolution allowed the creation of a mass market for electric lighting, and that it created the possibility of 'democratization of electric lighting'. Instead of tariffs based on low sales but high profit per unit of electricity sold, they argued for a tariff system based on lower profits per unit sold but greater sales.

The visionaries were proved right. Cheaper electricity combined with efficiency improvements created a mass market, much higher sales and greater profits. Electricity replaced gas for lighting, and the electricity revolution took off.

margin. The efficiency of electricity generation has improved ten-fold over the last century, and prices have fallen by over 90% (see also Book 1, Section 9.6 and Figure 12.9). The result is a mass market for electricity and the continual development of new electrical goods and services: electric lighting in the 1900s, domestic refrigeration in the 1930s; TV in the 1950s; microwaves and videos in the 1980s, computers and the internet in the 1990s.

Manufacturer and utility promotion

Manufacturers and utilities often work closely together to market new appliances, though their emphasis is on higher levels of energy services rather than energy savings. Box 5.6 describes the debate in the 1930s amongst manufacturers and utilities about the new fluorescent lighting. Utilities have long promoted energy efficiency as a useful aid to sales, particularly if there are competing fuels, as has been the case between electricity and gas in the heating and cooking markets.

The Electricity Council in the UK, through its research labs at Capenhurst, developed all-electric homes, built to much higher levels of insulation than required by the Building Regulations. Similarly they developed novel methods of industrial heating (e.g. infrared) and cooking (e.g. induction) to compete with gas appliances. The gas industry also conducted research into making its appliances more efficient and competitive with electricity, developing the condensing gas boiler, gas fired CHP etc. Even the coal industry developed more efficient coal boilers. However for success, high efficiency has to be combined with good performance, reliability and ease of use.

BOX 5.6 The introduction of fluorescent lighting in the 1930s

Some indication of the social and technical forces at work in the use of lighting can be seen in a study of the introduction of the fluorescent tube in the 1930s. Manufacturers were divided over whether to market this new type of bulb as a high-intensity or a high-efficiency lamp. Electric utilities expressed their concern over the loss of electricity sales, when they were faced with excess capacity. They prevailed on the manufacturers to drop the concept of high efficiency and instead to promote them as high intensity lamps, giving much greater levels of illumination for the same cost as existing filament bulbs.

The end result has been a very large increase in illumination levels in buildings that switched from filament to fluorescent lamps. For instance the designed illumination level for offices rose five-fold, from about 100 lux before 1940 to 500 lux by the 1970s.

Energy efficiency and economic growth

A major contributor to economic growth is greater productivity, or more efficient use, of the *factors of production* (such as capital, labour and energy), and this economic growth results in increased national energy consumption over time, though the linkages are not always clear (Schurr, 1990). Since increased energy efficiency is clearly a spur to economic growth, can we then argue that increased energy efficiency causes such a great increase in economic growth that energy consumption is *increased* - a rebound effect of greater than 100%? Alternatively, are we right to say that if there had been no energy efficiency improvements then energy consumption would have been higher? If so, then we could argue that energy efficiency improvements produced energy savings: that is, the difference between what would have occurred with and without the improvements. In neither case does there need to be an absolute reduction in energy consumption. These arguments rest on the notion that improvements in energy efficiency lower the rate of growth of energy consumption, and that the 'rebound effect' is less than 100% i.e. not all the energy savings are used up in the form of higher economic growth (Schipper, 2000). This ties in with the proposition that energy efficiency and economic growth are *not* closely linked, and that in fact economic growth can be 'decoupled' from energy consumption - the 'Factor 4' argument. (von Weizsacker *et al.*, 1997).

However these arguments are hypothetical, as are all macroeconomic arguments, since we cannot conduct economic experiments on society. Thus the hypothesis that energy efficiency improvements will lead to higher or lower national energy consumption (than would occur in the absence of such improvements) cannot be proven either way, anymore than the proposition that 'free trade leads to global economic growth' can be proved or disproved. All that can be demonstrated – through empirical research and theoretical argument – is that it is true (or untrue) in certain places, at certain times, or under certain conditions. It cannot be taken as a statement of what will definitely occur.

5.7 Critiques of energy efficiency

There have always been critics of the effectiveness of energy efficiency (as there always are of any aspect of government policy), but until the late 1990s in the UK they were confined to academic journals or the regional press. The first UK critic was Len Brookes, former Chief Economist at the UKAEA, who put forward the economists' viewpoint in *Energy and Environment* in 1990 (Brookes, 1990). Similar views were expressed in *The Scotsman* in June 1990 by John McCullough, a past chairman of the Institute of Energy in Scotland, who argued that there is an inescapable link between energy conservation, profitability and consumption: the effect of improved energy efficiency is increased consumption (Hancox, 1990).

Since then the debate over the effectiveness of energy efficiency has widened from critiques based on economics to those based on 'conservation ethics' (Rudin, 1999) and on a cultural and historical perspective (Moezzi, 1998). It also moved from the pages of academic journals (mainly *Energy Policy* and *Energy Journal*) to the more popular magazines such as *New*

Scientist with a critical article by Pearce (Pearce, 1998), which stimulated two special issues of academic journals (*Energy and Environment* and *Energy Policy*) in 2000 devoted to energy efficiency and the 'rebound effect'.

Critics such as Brookes pointed out that it might appear contradictory to expect increased consumption to come from improved efficiency. But Brookes (2000) argued that it is readily accepted in other areas of life that an improvement in efficiency (of a product) causes the (implicit) price of that product to fall, and hence stimulates consumption (see Section 5.6.1). For instance more efficient aircraft allow cheaper air fares, which result in a rise in air travel.

Environmentalists and efficiency

While many environmentalist accept that improved transport efficiency leads to increased mileage, few are willing to accept the similar case with energy use. One of the few is Duncan McLaren and colleagues who, in a Friends of the Earth book, comment:

> ...we need to be aware of the risk that we will exploit efficiency gains simply to consume more. It won't help having ...hyper-cars if we use them to drive around the corner instead of walking, and jet off around the world by plane at the drop of a hat! The real question is not so much 'how can we be that much more efficient', but 'how can we ensure the gains from our efficiency strategies are used to deliver real environmental improvements'?.
>
> McLaren *et al.*, 1998

Many environmentalists say that in the future there will be a shift in industrialized countries to 'dematerialization' of the economy, due to the structural shift in the economy from energy-intensive manufacturing to energy-frugal services (von Weizsacker, *et al.*, 1997). The debate on this matter is mostly beyond the scope of this chapter, but briefly, critics see flaws in this 'dematerialization' argument. For instance William Rees, a Canadian ecological economist, points out that improvements in the efficiency of resource use, such as those that can be achieved with computers, lead to a decline in the price of products, stimulating a mass market and hence a large global consumption of resources (i.e. the rebound effect). Rees also disputes the idea that moving from a primary (resource-based) or a secondary (manufacturing) economy to a tertiary, knowledge-based or service-oriented economy 'decouples' the economy from the environment (Wackernagel and Rees, 1997).

Similarly Ted Trainer, an Australian green activist, attacks the proposition that economic growth will be able to continue without increasing demands for materials and energy, commenting 'The essential element in this general position is a rejection of any need for fundamental change in lifestyles, culture or the economy' (Trainer, 1999). Thus 'deep green' advocates reject technical fixes, such as improving energy efficiency, as a means to reduce global environmental problems and instead stress the need to curb economic growth through adopting a low-consumption lifestyle, often termed 'voluntary simplicity' or a policy of 'sufficiency' (see Section 5.6).

Historical roots of the 'gospel of efficiency'

The advocacy of efficiency as the solution to resource shortages and environmental problems dates back to the early twentieth century. Samuel Hays, the American historian, in *Conservation and the Gospel of Efficiency* (first published in 1959) traces its origins back to the conservation policies that arose in the Progressive Era of 1890–1920 in the United States. As he argues:

> Conservation cannot be considered simply as a public policy, but far more significantly as an integral part of the evolution of the political structure of the modern United States.
>
> Hays, 1969

Furthermore, he writes:

> Conservationists were led by people who promoted 'rational' use of resources, with a focus on efficiency, planning for future use, and the application of expertise to broad national problems.
>
> Hays, 1969

Just as in the 1890s, when conservation policies were supported by large business organizations, so in the 1990s efficiency is still promoted by business interests, such as The World Business Council for Sustainable Development. Underlying such support is a system of decision-making about resource use based on expertise rather than political debate. As Hays remarks, it is a process:

> ...by which the expert would decide in terms of the most efficient dovetailing of all competing resource uses, according to criteria which were considered to be objective, rational, and above the give-and-take of political conflict. In short, they wanted to substitute one system of decision-making, that inherent in the spirit of modern science and technology, for another, that inherent in the give-and-take among lesser groupings of influence, freely competing within the larger system.
>
> Hays, 1969

Unlike the 'conservationists' of the 1970s, sometimes referred to as the 'prophets of doom', these nineteenth century conservationists were not prophets of despair. They displayed, according to Hays, 'that deep sense of hope which pervaded all those at the turn of the century, for whom science and technology were revealing visions of an abundant future' (Hays, 1969).

But this would not be a future negotiated through the normal processes of politics: it would be 'one guided by the ideal of efficiency and dominated by the technicians who could best determine how to achieve it'.

The ideology of energy efficiency

Hays' work on the ideology of efficiency has been used by two critics of American energy policy, Langdon Winner (1982) and Mithra Moezzi (1998). Langdon Winner, an American political scientist, commenting on the energy debate in the 1970s, wrote:

> Throughout the progressive era and in the decades since, an eagerness to define important public issues as questions of

> efficiency has been a common strategy...Thus it is not surprising to see efficiency reappear at the centre of today's energy debate. For Americans, to demonstrate the efficiency of a course of action conveys a sense of scientific truth, political wisdom, social consensus, and a compelling moral urgency.
>
> Winner, 1982

What interests Winner is the extent to which the prevailing terms of the energy debate have tended to focus exclusively on economic concepts like efficiency, and exclude other points of view, such as those based on equality and social justice. He remarks that the current debate among experts takes place within the framework of economic growth, in which efficiency, measured in terms of output per unit, is the only variable. This is because, he claims, energy analysts have simply been reflecting dominant attitudes in their society. He writes:

> ...growth - also called prosperity, abundance or progress - became a substitute for other shared ends. An unwavering faith in ever-expanding abundance became a way of ignoring some important questions about the shape of democratic society.
>
> Winner, 1982

Thus Winner argues that concentration on the concept of efficiency avoids questions about the structure of wealth and power in society. To avoid awkward political issues, he concludes, the policy with most consensual acceptance, 'is presumed to be that which seeks the most efficient use of "our" resources'.

Mithra Moezzi, an American energy analyst, observes that not only energy consumption, but also conservation and efficiency, have complex social meanings. She remarks:

> In the United States, the notion that efficiency is socially good and progressive dates at least to the beginning of the twentieth century...In contemporary American energy policy, the idea of progress through technical efficiency remains strong, though not without criticism.
>
> Moezzi, 1998

She argues that in the 1980s the US energy policy makers adopted the idea of energy efficiency through technology. This was in response to the tarnished image of energy conservation, which had left the US public sceptical, angry and alienated, with it being viewed as a sign of national weakness (Greenberger, 1983). She writes:

> The moralistic, anti-consumption tone of the energy conservation campaigns of the 1970s and early 1980s led the energy policy community to consciously turn away from a conservation focus toward an efficiency-orientated one.
>
> Moezzi, 1998

This shift in the 1980s tied in with the political shift towards 'free markets' and away from policies of government control over energy use. It also tied in with a backlash against the 1970s fears about 'running out of resources' and 'limits to growth' (Meadows, *et al.*, 1972; Simon and Kahn, 1984).

Ethical calls

Although most of the criticism of energy efficiency has come from the social sciences, there is a critique developing based on 'conservation ethics' as pioneered by Aldo Leopold (1949). Andrew Rudin, a long-standing critic of energy efficiency in the US, argues that what is needed for sustainability is not more efficiency (which he believes leads to greater consumption) but less consumption. Rudin firmly identifies efficiency with business values:

> Efficiency is good for business. Owners and investors emphasize that improved efficiency must be our mantra if America is to remain globally-competitive. Improved efficiency has also become the manifesto of our environment movement because the concept is politically correct, fundable and the basis of economic growth.
>
> Rudin, 1999

Rudin argues that improved efficiency rationalizes consumption by expanding the limits of natural resources, and justifies a wasteful lifestyle; instead there should be limits to consumption through moral restraint and cultural change. He therefore calls for a policy of conservation not efficiency, saying 'While efficiency tells us what to buy, conservation tells us how to behave' (Rudin, 1999). Furthermore he argues:

> ...if we want to protect the environment, we have to emphasize conservation and restraint, not improved energy efficiency and consumption. This is a moral issue, not an economic one...
>
> Conservation is heroic because it implies discipline, sacrifice, caring for common interests...We should use less energy because it is the right action, not just because someone pays us to do so.
>
> Rudin, 1999

Rudin advocates lifestyle changes: a return to the Jeffersonian ideal of the 'wisdom of frugality' (Thomas Jefferson was one of the 'Founding Fathers' of the USA and a supporter of a small scale agrarian society). He and others (see for example Douthwaite, 1996) advocate a 'deep green' society based on stable communities, local consumption and small-scale production by independent businesses.

However the conventional wisdom is that this sort of exhortation is doomed – just as it doomed President Carter. As the Royal Commission on Environmental Pollution (RCEP) comments:

> We consider that attempts to protect the environment and prevent climate change based principally on exhorting people to make sacrifices in comfort, pleasure and convenience in order to consume less energy are unlikely to succeed.
>
> RCEP, 2000

Policy implications

As Mithra Moezzi comments, promoting energy efficiency is not necessarily the best way to save energy or reduce pollution. Indeed, as she argues:

> It may actually encourage energy consumption by conveying the message that consuming increasing amounts of energy is acceptable as long as energy is consumed by technologies that have been deemed efficient.
>
> Moezzi, 1998

She believes that what is required is changes in consumer perception and behaviour, and that over-reliance on technical solutions is mistaken. For instance, she advocates that more emphasis should be given to absolute, rather than relative, energy consumption in energy labels and standards. A bigger refrigerator may be more 'efficient' but also consumes more than a smaller one. She remarks:

> The standards and guidelines inherent in energy policies reveal moral and technical judgements about what is 'good' and what is possible. When evaluating any particular standard, guideline, or other message related to energy conservation or efficiency, it is critical to consider not only 'energy saved' in a narrow sense, but what underlying messages are being conveyed and how they might affect the cultural perceptions of efficiency and consumption in short and long terms. By choosing specifications that reflect absolute rather than solely relative consumption, policies may encourage an ideological as well as practical shift in perceptions of energy use.
>
> Moezzi, 1998

Or in Rudin's homily: 'It's not the lamp, it's the switch'.

Sustainable consumption

Changes in behaviour and lifestyle have long been advocated as solutions to resource use problems. This desire by some environmentalists (and most religious teachers throughout the ages) to curb our material appetites has led to an upsurge of interest in the idea of 'sustainable consumption'. Laurie Michaelis, an Oxford researcher into the ethics of consumption, is well aware of the difficulties when he writes 'Modern conceptions of individual liberty and rights, property and just deserts, make it hard to imagine our society adopting controls on the type or volume of material consumption' (Michaelis, 2000a).

His studies emphasize the social importance of consumption and he believes that reductions are unlikely to be readily achievable on an individual basis. He believes that we should aim to develop ideals of the 'good life' that can be achieved without excessive material consumption. He concludes this is 'likely to require a cultural change ...deciding collectively how the good life should look, and to modify our behaviour accordingly' (Michaelis, 2000a).

However Finnish sociologists Eva Heiskanen and Mika Pantzar, while sympathetic to such moral changes, see difficulties. They argue that if anything has been learnt from consumer research on environmental issues, it is that changes in beliefs, attitudes or values do not necessarily lead to lifestyle changes. They comment:

> It is easy to agree that value change is needed, but new values are not swiftly taken up. Values are embedded in culture, both material and social... The dissemination of such ideas, and the setting in place of supporting institutions takes at least a hundred years. Obviously, we cannot wait that long for sustainable consumption.
>
> Heiskanen and Pantzar, 1997

Service efficiency

Society has generally preferred technical or economic solutions, and scientists and engineers have instead emphasized the technical possibilities of a shift to less resource-intensive types of consumption. One solution they advocate is the concept of 'service efficiency, which may be defined as providing a maximum of useful end-services to consumers using the minimum of materials and energy use' (Heiskanen and Pantzar, 1997). There is an extensive literature on this concept and there have been many attempts to design new types of service-producing machines that deliver energy services using innovative combinations of market goods and services and household labour. One such attempt is that by the Dutch Ministry of Environment in its programme *Sustainability and Quality Lifestyles for the Year 2000* which included such services as rent-a-car programmes, restaurant services, telecommunications and bulk delivery of goods to consumers.

Service efficiency in itself is not a panacea for sustainable consumption, as the gains are easily offset by an increase in the number and variety of products consumed. Also, as researchers into the cultural aspects of consumption have shown, it is necessary to understand how and why we consume. This is why innovations that are meant to be efficient and to reduce the need for resources often have the opposite effect.

As F-J Radermacher, a German sociologist, remarks:

> The trap that we have fallen into again and again over the course of technical progress consists of our always using progress on top of whatever went before (the rebound effect). This effect predicts that market forces and humanity's apparently unlimited capacity for consumption will use new technology to convert more and more resources into more and more activities, functions, services and products.
>
> Quoted in Hilty and Ruddy, 2000

It is often argued that information technologies, such as computers, will reduce material and energy consumption through substituting 'virtual' for real experiences and goods. But as Heiskanen and Pantzar ask:

> ...will the information super-highway do away with the urge to travel?...Will consumers actually substitute one good for another, or will they want to have it all: the television on, the newspaper on the table, and electronic news pointlessly self-scanning as the consumer of all this information dozes on the couch?
>
> Heiskanen and Pantzar, 1997

They make the comparison with claims made in the early 1980s about the paperless office, which never happened and actually turned out to be the opposite, with office paper consumption increasing.

Can we afford cost-saving energy efficiency?

This is the question posed by the ecological economists William Rees and Mathis Wackernagel. They write:

> The answer is 'yes' only if efficiency gains are taxed away or otherwise removed from further economic circulation. Preferably they should be captured for reinvestment in natural capital rehabilitation.
>
> Wackernagel and Rees, 1997

In other words they propose environmental taxes, and there has been much work done by ecological economists, such as Robert Costanza and his colleagues in the US (1996) and Paul Ekins in Britain (1999), on 'ecotaxation' - that is, shifting the tax burden from employment to (non-renewable) resources, particularly energy. Some classical economists however disagree with this approach on conservation of resources. They argue that the best legacy one can bequeath to the next generation is a high level of real output per capita, and voluntarily foregoing opportunities to that end for mistakenly altruistic reasons may be doing a disservice to the next generation. The choice of whether to bequeath natural or (man-made) capital goods to future generations lies at the heart of the debate on 'sustainability', and on 'soft' (light green) versus 'hard' (deep green) approaches to achieving it.

One feasible way to cut energy consumption is through energy taxes and regulation, but these involve economic costs to society and are not politically popular in the UK –particularly with motorists. Consumers seem unwilling to pay a higher price for 'green' electricity - that is, electricity generated from renewable energy. Only 45 000 households (about 3 in 1000 of UK households) had signed up by March 2002 for 'green power' schemes offered by utilities, compared to 20% of households in the Netherlands. For further details see www.greenprice.com [accessed 7 June 2003].

Combining efficiency and green power

One way to stimulate the market might be a combined policy of 'green' electricity and energy efficiency. This could be done if energy efficiency packages were sold along with 'green' electricity by utility companies. The cash savings from the efficiency measures would be used to off-set the higher cost to consumers of 'green' electricity. As Elliott remarks:

> It strikes me that the marketing process could be enhanced if the purveyors of 'white goods', such as washing machines or tumble dryers, could offer their customers an opportunity to switch over to green power when they buy their new consumer appliance. I can image the salesperson saying, 'fine, now you have your nice new fridge freezer, how are you going to power it? Surely not on dirty old fossil fuels or nuclear power? Why not buy some of this nice clean green power, so that you won't be adding to the world's environmental problems'.
>
> Elliott, 2001

Offering green power at the point of sale – whether at the shop or on the Web – makes a lot of sense environmentally as well as commercially. Offering green-power packages is like selling internet access packages when you buy a computer. It could even be extended beyond white goods and consumer electronics to central-heating systems and even to electric cars. It might be possible to offer green-power vouchers instead of cash refunds when you buy the most efficient appliance. As Elliott continues:

> ...it is important that cash savings from the installation of energy conservation measures are fed back to support green power, rather than being re-spent on energy intensive goods and services. Otherwise the re-spent cash could involve consumption of conventional power, and could increase, rather than reduce, net greenhouse emissions.
>
> Elliott, 2001

Government action

But green consumers and retailers can only do so much. Governments need to act, either through regulation or taxation, if we are to achieve a 60% reduction in greenhouse gas emissions (as indicated by the RCEP) or even the 20% cut by 2010 aimed for by the current Government. However, as the OECD warns in its report on 'eco-efficiency', there is a danger in making energy efficiency, basically an indicator of energy productivity, a goal of public policy. It remarks that government policies:

> ...are best aimed at improving the state of the environment, the health of the economy and the quality of life, rather than increasing productivity *per se*. Where efficiency improvements alone are sought, there is a risk that increases in economic activity will outweigh any reductions in environmental impact per unit of activity.
>
> OECD, 1998

It is important, it counsels, to be aware that regulations and voluntary agreements to increase resource efficiency can have unwanted environmental impacts. Thus the OECD report argues that efficiency standards should be seen as complements rather than alternatives to eco-taxes. These two instruments have different economic rationales, and should be applied together.

The climate change levy

The only recent fiscal action in the UK aimed at reducing carbon emissions has been the introduction of the Climate Change Levy, which is a tax on all energy use outside the domestic and transport sectors that came into effect from April 2001. The Levy is designed to be 'revenue neutral' with refunds in the form of lower National Insurance contributions from employers, as well as extensive rebates for energy-intensive users. As originally proposed, the Levy was purely an energy tax, so there was no incentive to switch to carbon-free sources. However after consultations it was transformed into a carbon tax of sorts - with renewables (but not nuclear power) exempt from the Levy.

What role for energy efficiency

In the UK over the past 50 years, primary energy consumption has approximately doubled despite great improvements in energy efficiency (see Book 1, Figure 2.8). This suggests that we have preferred to take much of the efficiency gains in the form of higher levels of energy service, such as using central heating, rather than reduced consumption. This is not surprising since the effect of increased efficiency is to lower the implicit price of an energy service, and hence make its use more affordable to existing and new consumers.

The history of 'efficiency' policy is a long one, dating back a century, but is intimately bound up with notions of progress, growth and power. There is indeed an unending race between energy efficiency and economic growth. If growth is faster than the rate of efficiency increase (as it has been, historically) then total energy consumption increases. For instance in the UK between 1965 and 1998 'efficiency', as expressed by energy intensity (primary energy per unit of GDP), roughly doubled - a Factor 2 improvement. However GDP *more than* doubled: it rose by 147%, so total energy consumption rose by a quarter. Thus at current rates of efficiency improvement, it is perfectly feasible for there to be a Factor 4 improvement in the course of this century. But as the RCEP comments:

> There will continue to be very large gains in energy and resource efficiency, but on current trends we find no reason to believe that these improvements can counteract the tendency for energy consumption to grow. Even if energy consumed per unit of output were reduced by three-quarters or Factor Four, half a century of economic growth at 3% a year (slightly less than the global trend for the past quarter century) would more than quadruple output, leaving overall energy consumption unchanged.
>
> RCEP, 2000

The race between growth and efficiency

A key debate about energy efficiency is about the extent of the 'rebound': to what extent does energy use increase due to efficiency improvements? Does greater efficiency lead to higher or lower energy use than there would have been without those improvements? This is at heart a hypothetical question. Thus this debate, like most economic questions, can never be resolved: there is only theoretical argument and analysis of past trends (Wirl, 1997; Saunders, 2000; Schipper, 2000). But one thing does seem clear: promoting energy efficiency alone is unlikely to lead to an absolute reduction in energy consumption. Efficiency will always lose the race against economic growth, unless there are constraints on consumption – through taxation, regulation or changes in lifestyle.

I believe that the current Government emphasis on energy (or eco-) efficiency as a technical means to reduce CO_2 emissions is misguided. As Roger Levett observes in his chapter on 'eco-efficiency' in a report to the UK Sustainable Development Commission:

> There is absolutely no evidence from past trends that improvements in eco-efficiency can achieve the 60% reduction in energy use and/ or greenhouse emissions that the IPCC has called for as a precautionary first step.
>
> Levett, 2001

Levett argues instead for a shift towards reducing consumption through behavioural changes, such as car sharing or walking.

I conclude that like so many 'technical fixes', energy efficiency improvement is a flawed solution to the problem of global warming. As Laurie Michaelis remarks:

> ...even factor 4 improvements over 30 years would be very difficult to achieve without substantial changes in government policy and consumer choice. Nor can it be guaranteed that improving resource efficiency in the industrialised countries will reduce resource use by the wealthy. In the long run, their perceived needs and demands for goods and services may grow to take up the slack in resource availability. This can happen either because reduced production costs stimulate demand, or because innovation to improve

> efficiency leads to the generation of new products and new areas of consumer demand.
>
> Michaelis, 2000b

Michaelis thus argues for changing consumption patterns through a combination of technological and social changes.

Deep green solutions

Some environmentalists (e.g. Carley and Spapens, 1998) readily accept that merely increasing efficiency does not lead to reduced consumption. They argue for the need to de-link economic growth from resource consumption, and for a policy of 'sufficiency', which is 'living well on less'. Such a policy of 'sufficiency', is also advocated by deep greens like Sachs (1999), Trainer (1995) and Douthwaite (1996), but involves a massive challenge to the present dominant ideology of free market capitalism. In their efforts to combat global warming, most Western governments eschew such politically-controversial solutions and instead seek technical fixes, such as energy efficiency, to solve the problem.

However promoting efficiency without curbs on consumption (through regulation or taxation) will not solve the problem of reducing CO_2 emissions. The goal should be lower carbon dioxide emissions, not lower energy use; ultimately, energy growth needs to be decoupled from CO_2 emissions. In the long term, a shift to a world economy entirely powered by zero-carbon fuels such as renewables could in principle allow continued energy growth with virtually no carbon emissions, enabling economic growth to continue indefinitely – though whether this would be a good thing is another matter (see Boyle, 2000). Thus the emphasis should be on switching to non-fossil fuels, such as renewables, which are a better 'technical fix' than energy efficiency measures, although governments should still promote energy efficiency on both economic and social-welfare grounds. Eco-taxation, such as a carbon tax, could be used to subsidize the introduction of non-fossil fuels. The aim of such taxation should not be prescriptive, to get consumers to use less energy, but to raise revenue.

What is the 'good life'

The key questions we have discussed in this Chapter are ethical not technical, cultural rather than economic. What is the 'good life'? Can we consume more goods and services (for a higher quality of life) but use less materials and energy? Can a low(er) energy lifestyle be made desirable by moral suasion or cultural example?

These are age-old questions: achieving a consensus on them and developing practical ways of moving towards a 'conservation' lifestyle will take time. In the meantime, energy efficiency is a valuable tool to save consumers' money and stimulate economic productivity, and it should still be promoted whatever its impact on energy consumption.

However I believe that ultimately what is needed if we want to limit the damaging environmental impact of energy production, is a policy of energy 'sufficiency' - that is, living well on less energy. In the meantime we need somehow to de-link economic growth from energy consumption. For both of these strategies, energy efficiency is a key tool, but the end result will depend on how we use it.

References and data sources

Bernardini, O. and Galli, R. (1993) 'Dematerialization: Long Term Trends in the Intensity of Use of Materials and Energy', *Futures*, May 1993, pp. 431–448.

Bijker, W. (1995) *Of Bicycles, Bakelite and Bulbs: Towards a Theory of Technological Change*, MIT Press.

Boulding, K. (1966) 'The Economics of the Coming Spaceship Earth' in Jarrett H. (ed.) *Environmental Quality in a Growing Economy*, Baltimore, John Hopkins University Press.

Boyle, G. (2000) 'DREAM-World: a simple model of energy related carbon emissions in the 20th and 21st centuries', *Energy and Environment*, vol. 11, no. 5 pp. 573–86.

Brookes, L. (1990) 'Energy Efficiency and the Greenhouse Effect', *Energy and Environment*, vol. 1, no. 4 pp. 318–30.

Brookes, L. (2000) 'Energy efficiency fallacies revisited', *Energy Policy*, vol. 28 no. 6–7, pp. 355–66.

Carley, M. and Spapens, P. (1998) *Sharing the World: Sustainable Living and Global Equity in the 21st Century*, Earthscan.

Chapman, J. and Eyre, N. (2001) *Energy Productivity to 2010 – Potential and Key Issues*, Performance and Innovation Unit, UK Cabinet Office, http://www.cabinet-office.gov.uk/innovation/2002/energy/report/working%20papers/PIUd.pdf [accessed 13 May 2003].

Costanza, R., Segura, O. and Martinez-Allier, J. (1996) *Getting Down to Earth: Application of Ecological Economics*, Island Press.

Daly, H. and Cobb, J. (1989) *For the Common Good: Redirecting the economy towards community, the environment, and a sustainable future*, Boston, Beacon.

de Beer, J. (1998) *Potential for Industrial Energy Efficiency Improvement in the Long Term*, The Netherlands, University of Utrecht.

DEFRA (1999) *Fuel Poverty: The New HEES – a programme for warmer, healthier homes*, Department of Environment, Food and the Regions. http://www.defra.gov.uk/environment/consult/fp/index.htm [accessed 13 May 2003].

Douthwaite, Richard (1996) *Short Circuit*, Green Books.

DTI (2003) *Our Energy Future – creating a low carbon economy*, Department of Trade and Industry, http://www.dti.gov.uk/energy/whitepaper/ourenergyfuture.pdf [accessed 13 May 2003].

Ekins, P. (1999) *Economic Growth and Environmental Sustainability*, Routledge.

Elliott, D. (2001) 'Can Green Consumers take over Power?', *Renew* No. 130. March/April 2000, pp. 18–21.

Greene, D., Kahn, J. and Gibson, R. (1999) 'Fuel Economy Rebound Effects for US Household Vehicles', *The Energy Journal* vol. 20, no. 3, pp. 1–29.

Greening, L., Greene, D. and Difiglio, C. (2000) 'Energy efficiency and consumption - the rebound effect - a survey', *Energy Policy*, vol. 28 nos 6–7, pp. 389–401.

Greenberger, M. (1983) *Caught Unawares: The Energy Decade in Retrospect*, Ballinger Publishing Group.

Hancox, J. (1990) 'Stoking the fires of consumerism', *The Scotsman*, 12 June 1990.

Hawken, P., Lovins, A. and Lovins, H. (1999) *Natural Capitalism: The Next Industrial Revolution*, Earthscan.

Hays, S. P. (1969) *Conservation and the Gospel of Efficiency: the Progressive Conservation Movement, 1890–1920*, Harvard University Press.

Heiskanen, E. and Pantzar, M. (1997), Towards Sustainable Consumption: Two New Perspectives, *Journal of Consumer Policy* vol. 20 pp. 409–442.

Herring, H. (2000) 'Is Energy Efficiency Environmentally Friendly?', *Energy and Environment*, vol. 11, no. 3. pp. 313–326.

Hilgartner, S., Bell, R. and O'Connor, R. (1982) *Nukespeak: Nuclear Language, Visions And Mindset*, Sierra Club Books, San Francisco.

Hilty, L. and Ruddy, T. (2000) 'Towards a Sustainable Information Society', *Informatik* no. 4, August 2000, pp. 2–9.

IEA (1997) *Indicators of Energy Use and Efficiency: Understanding the link between energy and human activity*, International Energy Agency, Paris.

Leopold, A. (1949) *A Sand Country Almanac*, Oxford University Press.

Levett, R. (2001) *State of sustainable development in the UK*, Report to the Sustainable Development Commission, CAG Consultants, London, Feb. 2001.

McLaren, D., Bullock, S. and Yousuf, N. (1998) *Tomorrow's World: Britain's Share in a Sustainable Future*, Earthscan.

Meadows, D. *et al.*, 1972, The Limits to Growth, Pan Books.

Michaelis, L. (2000a) *Ethics of Consumption*, Centre for Environment, Ethics and Society, Oxford.

Michaelis, L. (2000b) *Sustainable Consumption: A Research Agenda*, Centre for Environment, Ethics and Society, Oxford.

Moezzi, M. (1998) *The Predicament of Efficiency*, Proceedings of the 1998 ACEEE Summer Study on Energy Efficiency in Buildings, August 1998. pp. 4.273–4.282, http://enduse.lbl.gov/Info/Pred–abstract.html [accessed 13 May 2003].

Moezzi, M. (2000) 'Decoupling energy efficiency from energy consumption', *Energy and Environment*, vol. 11, no. 5 pp. 521–538.

Milne, G. and Boardman, B. (2000) 'Making cold homes warmer: the effect of energy efficiency improvements in low-income homes', *Energy Policy*, vol. 28 no. 6–7, pp. 411–424.

NAO (2003) *Six out of ten could get better deal on electricity bills*, Press release, National Audit Office, 9 May 2003, http://www.ofgem.gov.uk/temp/ofgem/ cache/cmsattach/3169 r4003 9may.pdf [accessed 14 May 2003].

OECD (1998) *Eco-Efficiency*, OECD, Paris.

Owen, G. (1999) *Public purpose or private benefit? The politics of energy conservation*, Manchester University Press.

Pearce, F. (1998) 'Consuming Myths', *New Scientist,* 5 September 1998, pp. 18–19.

PIU (2001) *Energy Efficiency Strategy*, Performance and Innovation Unit, UK Cabinet Office, http://www.cabinet-office.gov.uk/innovation/2002/energy/report/working%20papers/PIUc.pdf [accessed 13 May 2003]

RCEP (2000) *Energy – The Changing Climate.*, Stationery Office, London, http://www.rcep.org.uk/energy.html [accessed 13 May 2003]

Rudin, A. (1999) *How improved efficiency harms the environment*, http://home.earthlink.net/~andrewrudin/article.html [accessed 13 May 2003]

Sachs, W. (1999) *Planet Dialectics*, Zed Books.

Saunders, H. (2000) 'A view from the macro side: rebound, backfire, and Khazzoom-Brookes', *Energy Policy,* Vol. 28 no. 6–7, pp. 439–449.

Schipper, L. and Meyers, S. (1992) *Energy Efficiency and Human Activity: Past Trends, Future Prospects*, Cambridge University Press.

Schipper, L. (2000) 'On the rebound: the interaction of energy efficiency, energy use and economic activity: An introduction', *Energy Policy*, vol. 28, no. 6–7, pp. 351–53.

Schurr, S. H. (1990) *Electricity in the American Economy: Agent of Technological Change*, Greenwood Press, USA.

Simon, J. and Kahn, H. (1984) *The Resourceful Earth: A Response to Global 2000*, Basil Blackwell.

Sorrell *et al.,* (2000) *Barriers to energy efficiency in public and private organisations*, SPRU, University of Sussex, http://www.sussex.ac.uk/spru/environment/research/barriers.html [accessed 13 May 2003]

Trainer, T. (1995) *The Conserver Society: Alternatives for Sustainability*, Zed Books.

Trainer, T. (1999) 'Exploring Energy Solutions for Industrial Society', *Energy and Environment* vol. 10, no. 3 pp. 293–304.

Wackernagel, M. and Rees, W. (1997) *Our Ecological Footprint: Reducing Human Impact on the Earth*, Gabriola Island, B.C., Canada, New Society Publishers.

von Weizsacker, E., Lovins, A. and Lovins, H. (1997) *Factor four: doubling wealth – halving resource use*, Earthscan.

Winner, L. (1982) 'Energy Regimes and the Ideology of Efficiency' in Daniels, G. H. and Rose, M. (eds), *Transport and Energy: Historical Perspectives on Contemporary Policy*, Sage, pp. 261–77.

Wirl, F. (1997) *The Economics of Conservation Programs*, Kluwer Academic.

WRI (1997) *Resource Flows: The Material Basis of Industrial Economies*, World Resource Institute, Washington D.C.

Further Reading

Binswanger, M. (2001) 'Technological Progress and Sustainable Development: What about the Rebound Effect?', *Ecological Economics* vol. 36, pp. 119–132.

DTI (2000), *Energy Paper 68: Energy projections for the UK*, Department of Trade and Industry. http://www.dti.gov.uk/energy/ inform/ energy_projections/index.shtml [accessed 13 May 2003]

Herring, H. (1998) *Does Improving Energy Efficiency save Energy: the Economist Debate*, EERU Report No. 74, the Open University, August 1998, http://www-tec.open.ac.uk/eeru/staff/horace/hh3.htm [accessed 13 May 2003]

Nye, D. (1998) *Consuming Power: a Social History of American Energies*, MIT Press.

Rudin, A. (2000) 'Let's stop wasting energy on efficiency programmes', *Energy and Environment*, vol. 11, no. 5, pp. 539–552.

Sanne, C. (2002) 'Willing consumers – or locked-in? Policies for a sustainable consumption', *Ecological Economics*, vol. 42, no. 1–2, pp. 273–87.

Sorrell, S. (2002) 'Making the link: climate policy and the reform of the UK construction industry', *Energy Policy*, vol. 31, pp. 865–78.

Sutherland, R. (2000) '"No Cost" Efforts to Reduce Carbon Emissions in the U.S.: An Economic Perspective', *Energy Journal*, vol. 21, no. 3, pp. 89–112.

Wernick, I., Herman, R., Govindi, S. and Ausubel, J. (1996) 'Materialization and Dematerialization: measures and trends', *Daedalus*, 125, pp. 171–98.

Wollard, R. and Ostry, A. (2000) *Fatal consumption: rethinking sustainable development*, UBC Press, Vancouver, Canada.

Acknowledgements

Grateful acknowledgement is made to the following sources for permission to reproduce material within this book:

Chapter 1

Figures

Figures 1.1,1.5a,1.8 and page 23: Courtesy of Stephen Potter; Figure 1.3: Press Association/Batchelo Barry Batchelor; Figure 1.5b: © G. Brad Lewis/ Science Photo Library; Figure 1.6: Courtesy of John Hopkins University Applied Physics Laboratory.

Chapter 2

Text

Box 2.8: Wainwright, M. 'Chicken fat to power lorries', *The Guardian,* 28 October, 2002, © The Guardian; Box 2.7: Adapted from *Case Study: LPG Vehicles: Oxfordshire Mental Healthcare NHS Trust*, The Energy Saving Trust; Box 2.11: Adapted from *Case Study: Electric Vehicles Nottingham Millennium Electric Vehicle Project*, The Energy Saving Trust.

Figures

Figure 2.1: © National Motoring Museum; Figure 2.2: © David Noble Photography; Figures 2.4 and 2.6: © P. M. Northwood; Figures 2.9 and 2.11: © Science Photo Library; Figure 2.10: © Ecoscene/ Joel Creed; Figure 2.12: Courtesy of Daimler-Chrysler.

Chapter 3

Text

Box 3.4: Extracts from 'Case Study: Plymouth Hospitals NHS Trust', *Making Travel Plans Work: Case Study Summaries*, July 2002, Department of Transport. Crown copyright material is reproduced under Class Licence Number C01W0000065 with the permission of the Controller of HMSO and the Queen's Printer for Scotland. Box 3.5: Extracts from 'Case Study: Nottingham City Hospital NHS Trust', *Making Travel Plans Work: Case Study Summaries*, July 2002, Department of Transport. Crown copyright material is reproduced under Class Licence Number C01W0000065 with the permission of the Controller of HMSO and the Queen's Printer for Scotland.

Figures

Figures 3.1–3.6, 3.8(b), 3.9: © Marcus Enoch; Figure 3.7: Local Transport Today/Nick Jeanes; Figure 3.8(a): © Sally Cairns; Figure 3.8(c) © Stephen Potter.

Chapter 4

Text

Box 4.1: Extracts from 'Commuter Plans in Cheshire: steps to success', www.cheshire.gov.uk, courtesy of Cheshire County Council; Box 4.2: Extracts from 'Roles and Responsibilities', *A Travel Plan Resources Pack for Employers*, 2000, The Energy Saving Trust; Box 4.3: Extracts from '*Good Practice Case Study 402: Heathrow Airport Retail Consolidation Centre*', Energy Efficiency Best Practice Programme, Crown copyright material is reproduced under Class Licence Number C01W0000065 with the permission of the Controller of HMSO and the Queen's Printer for Scotland.

Figures

Figures 4.1, 4.2, 4.4, 4.5, 4.8(a) and 4.9: © Marcus Enoch; Figures 4.3 and 4.6: © Sally Cairns; Figure 4.8(b): Cranfield University Press.

Chapter 5

Text

Various text and table 1: *Energy Efficiency Strategy,* Performance and Innovation Unit, UK Cabinet Office. Crown copyright material is reproduced with permission of the Controller of HMSO.

Figures

Figures 5.1 and 5.3: 'Save it', mid 1970s and 'Monergy' late 1970s. Crown copyright material is reproduced with permission of the Controller of HMSO; Figure 5.4: de Beer, J. (1998) Potential for Industrial Energy Efficiency Improvement in the Long Term, University of Utrecht.